The DNA Detectives

The DNA Detectives

How the Double Helix Is Solving Puzzles of the Past

Anna Meyer

THUNDER'S MOUTH PRESS
NEW YORK

For my grandpa,
Alexander Park Gregg
A true gentleman scientist

THE DNA DETECTIVES
HOW THE DOUBLE HELIX IS SOLVING PUZZLES OF THE PAST

Published by
Thunder's Mouth Press
An Imprint of Avalon Publishing Group, Inc.
245 West 17th Street, 11th Floor
New York, NY 10011

Originally published as *Hunting the Double Helix* in Australia in 2005 by Allen & Unwin
First Thunder's Mouth Press edition April 2006

Diagrams on pages 3, 20, 105, and 163 are by Ian Faulkner.

Library of Congress Cataloging-in-Publication Data is available.

ISBN: 1-56025-863-2
ISBN 13: 978-1-56025-863-6

9 8 7 6 5 4 3 2 1

Book design by Midland Typesetters
Printed in the United States of America
Distributed by Publishers Group West

CONTENTS

Acknowledgments vii
Introduction: *The quest begins* 1

1 Almost Human: *Were the Neanderthals our ancestors?* 9
2 The Prehistoric Zoo: *Could extinct animals be brought back to life?* 36
3 Cretaceous Capers: *Could we really clone a dinosaur?* 66
4 Big Bird: *Unravelling the mysteries of the New Zealand moa* 91
5 Plague Proportions: *Searching for the truth behind devastating historical epidemics* 113
6 Problems of Identity: *Did Anastasia survive the Russian Revolution?* 144
7 The Heart of the Matter: *What became of Louis XVII of France?* 173

Conclusion: *What next for ancient DNA research?* 194
Sources 199
Index 225

ACKNOWLEDGMENTS

First, I sincerely thank everyone at Allen & Unwin who has played a part in this book, especially Ian Bowring, Emma Cotter and Angela Handley. The fantastic advice and continuous support you have given me has been absolutely invaluable.

I greatly appreciate the assistance of the experts in ancient DNA who have generously provided information and read drafts of chapters. In particular, I would like to thank Jean-Jacques Cassiman, Adrian Gibbs, Arthur Aufderheide and David Lambert.

A big thank you to my PhD supervisors at the Australian National University's Centre for the Public Awareness of Science, Sue Stocklmayer and Chris Bryant, for your support and input into this project. I also acknowledge with great appreciation the support of an ANU PhD scholarship while working on this project.

My deepest thanks go to my husband, Andrew Dickson, for his unwavering support and encouragement, reading endless drafts and, most of all, believing in me. Thank you also to the rest of my family for providing plenty of support and encouragement—no one could wish for more. Finally, thank you to my dog Sweep, my constant and faithful companion through this entire process.

INTRODUCTION
The quest begins

One morning a few years back, I was in my university office, busily photocopying a stack of boring scientific articles. Hearing a sound outside the room, I glanced up and found myself confronted with rather an unusual sight. Striding down the corridor towards me at quite a pace was a tall, thin man, with unruly shoulder-length brown hair and an eager, slightly wild-eyed expression. In his hand he clutched a small plastic bag, containing one small, slightly squashed, very dead bird. 'It's a bird in a bag!' he announced enthusiastically, stating the obvious, before turning the corner and disappearing into the genetics laboratory.

'Who was that?' I thought to myself. I was intrigued by this bird-toting stranger, and I was curious about exactly what he intended to do in a genetics laboratory with a squashed bird. It was by far the most interesting thing I had seen all day. I asked my supervisor, David, who he was. 'That's Alan Cooper,' David replied. 'He does ancient DNA research at Oxford University, and he's over here doing some experiments on that bird today.' 'Oh, right,'

I replied, none the wiser—I didn't have a clue what ancient DNA research was, but I didn't want to admit it.

Now at the time I was looking around for a topic for an essay I had to write for one of my postgraduate genetics courses. I had been thinking about it for ages, but no ideas had sprung to mind. I didn't know anything about ancient DNA but it sounded interesting, so I thought, 'Why not?' and decided to write my essay on it.

Seizing the opportunity, I cornered Alan on his way back down the corridor, and asked if he would tell me a bit about ancient DNA research. 'I'm just heading over to buy some lunch,' he said, 'follow me and I'll tell you about it.' 'Great!' I replied. Over a hasty lunch, Alan explained the basics of ancient DNA research and I was immediately hooked. It's actually really awesome.

Basics of DNA

Before I explain about ancient DNA, here's a brief overview of a few important facts to do with DNA. Humans, and in fact living things in general, are in effect nothing more than massive collections of almost unimaginably large numbers of discrete microscopic cells, all working cooperatively to produce a unique individual.

Inside virtually every cell is a specialised compartment, the nucleus, which contains a set of small dense structures, the chromosomes. Each species has a characteristic number of chromosomes per cell—humans, for example, have 46.

Opposite: *The cell nucleus contains a number of chromosomes, each consisting of a long, tightly wound DNA molecule. DNA exists in the form of a 'double helix', with two paired strands wound around each other. A strand consists of a string of four bases: adenine* (A), *thymine* (T), *guanine* (G) *and cytosine* (C). *In the double helix,* A *always pairs with* T, *and* G *always pairs with* C. *DNA can be obtained from virtually all cells.*

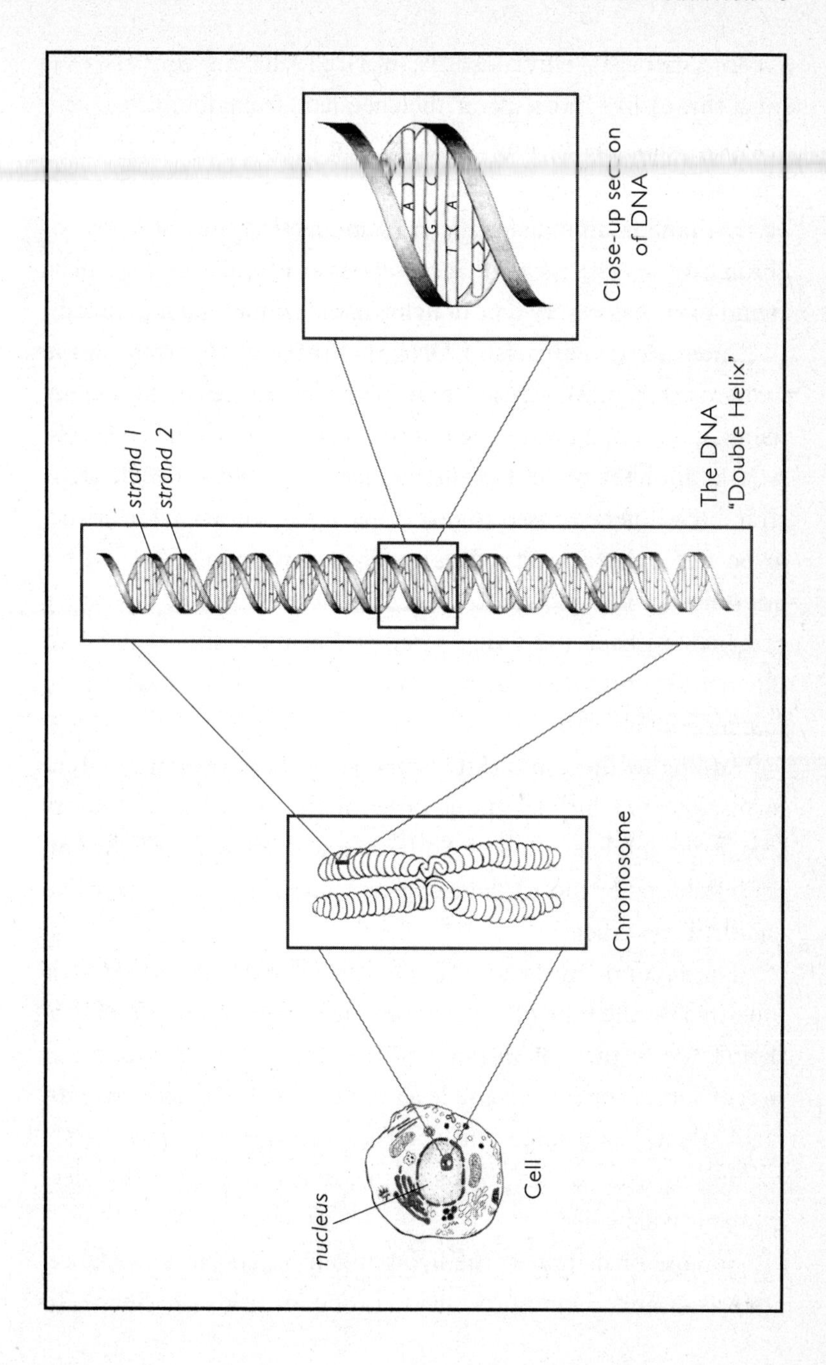
nucleus
Cell
Chromosome
strand 1
strand 2
The DNA
"Double Helix"
Close-up sec on
of DNA

In turn, each chromosome in a cell nucleus consists of a long, thread-like molecule of the chemical deoxyribonucleic acid (DNA), which is wound up and bundled tightly to give the chromosome its characteristic shape.

Each cell in an individual carries an identical copy of the set of chromosomes. This means that copies of an individual's DNA can be found in virtually every type of living tissue within that individual.

There are two aspects of DNA that make it important in the living organism. The first is that DNA determines or influences many aspects of how an organism looks, functions and behaves. A large number of units called genes are found on each DNA strand. Each gene is used to determine a particular characteristic, or part of a characteristic. It is a rather complex process, but for our purposes it is enough to know that there are tens of thousands of genes on each individual's set of chromosomes, which can interact in a multitude of ways to influence different aspects of that individual.

Adding to the complexity, some genes have several different forms, each of which leads to a variation in a particular characteristic, for example eye colour. Some characteristics, such as height, are influenced by more than one gene, as well as by environmental factors such as diet.

The second important characteristic of DNA is that it is inherited. At the time of conception, each parent passes on part of their DNA to their offspring, and it is through this mechanism that offspring come to resemble a unique mixture of both parents. DNA, then, has a huge and complex involvement in not only making us what we are, but also in determining what our descendants will be like.

Because of its role in the living organism and in inheritance, DNA is of great interest in a wide variety of fields of science. To name a few, the study of DNA can be used to examine aspects of how living organisms function, the evolutionary relationships between species, how inherited diseases occur and, more recently, the production of genetically modified organisms.

Extracting DNA

Obtaining DNA from living organisms for study is a fairly straightforward process, and in essence it is the same for any biological material. First a sample from the organism of interest is obtained. Because of its ubiquity within living tissues, DNA can be extracted from virtually any part of a plant or animal, including skin, hair, blood, bone, teeth, seeds, leaves, insects, fungi and even bacterial colonies. The list goes on. The sample is ground up to separate all the cells, and chemicals—basically detergent, enzymes and alcohol—are added, which separate the DNA from the other parts of the sample. After this process, the DNA sits in a test tube, in a relatively pure form, ready for further study. Strangely enough for such an important molecule, pure DNA is not actually very spectacular to look at: it is a whitish, thread-like material, which can be scooped up and wound around a glass rod—a common, if rather nerdy, source of entertainment for undergraduate genetics students in their laboratory classes.

What happens to the extracted DNA next depends on the aim of each individual experiment. Sometimes the next step involves using enzymes to chop the DNA into large but manageable pieces, which can then be further analysed. Usually, however, only a small part of the organism's entire genome (the full set of DNA contained on all the chromosomes in a cell) is of interest in any particular study, and so copies are made of these parts (using a process I will explain more fully later on). For example, copies may be made of one or more individual genes, which effectively separates them from the rest of the DNA. These copied sections can then be studied in detail; for example, they can be compared with the equivalent sections of DNA from a different species, to get an idea of how closely related the two species are, or they can be used to learn more about how individual genes function.

Ancient DNA

The extraction and study of DNA from living tissues is a well-established scientific discipline. However, it was discovered only recently that DNA can be found not only in living things, but also in the remains of organisms *that are no longer alive.* It was not long before this remarkable discovery led to the establishment of an entirely new scientific discipline—the study of ancient DNA.

The field of ancient DNA research involves the study of any DNA that still exists in the remains of once-living organisms. Like a window into the distant past, ancient DNA has been found in a whole variety of organisms that have been dead for anything from around 100 years—such as the extinct Australian thylacine (commonly known as the Tasmanian tiger) and the New Zealand moa—right up to tens of thousands of years, for example, the Neanderthals and woolly mammoths.

The fact that DNA can survive for such a long time is, in itself, pretty surprising. But there is much more to it than that. By studying ancient DNA in detail, the most amazing discoveries have been made. Ancient DNA is a relatively recent technology, with research in the field only beginning in the mid-1980s. Nonetheless, it has already been involved in a whole smorgasbord of delicious stories. There are tales of murder, deadly disease, mysterious disappearances, animals that have long been extinct, and even discoveries about human origins.

Thanks to Alan, I had more than enough material for my essay. Alan went back to his ancient DNA lab at Oxford University, and I almost forgot about ancient DNA for a few years.

Almost, but not completely. The concept had me so fascinated that it stayed at the back of my mind somewhere, and I couldn't help thinking about it from time to time. It didn't take a lot of agonising to decide to write a whole book on ancient DNA when the opportunity arose a couple of years ago.

In the chapters ahead, I have included the stories behind some of the most interesting ancient DNA discoveries that have been made so far. I hope that by the time you finish reading them, you will be as hooked on ancient DNA as I am.

1

ALMOST HUMAN

Were the Neanderthals our ancestors?

While I was writing this chapter, I made the curious discovery that you can live with someone for years before you notice something about them that is actually quite striking. I suppose it is because they become so familiar that you stop looking *really* closely at them. Then, for some reason, something makes you notice that striking feature, and from then on you can't stop fixating on it.

The other night, I was sitting on the couch next to my husband, Andrew. We were watching television, our dog snoring away happily at our feet, paws twitching as he chased small furry animals in his sleep. A commercial break began and I turned to Andrew to ask him something. It was then that it happened: I noticed that my husband—whom I have known for a large part of my life, and seen virtually every day since I met him—has *enormous* eyebrow ridges.

When I run my hand over my forehead, from my hairline down over my eyebrows, my skull feels smooth, with just the slightest mound of bone pushing outwards at the top of my eye sockets where my eyebrows are. Andrew's eyebrows, on the other hand, sit on a ridge of bone that juts out over his eyes like an overhanging cliff face. His brow ridges are so large they actually cast shadows over his eyes and, in the right light, down onto his cheeks. I had never noticed these monolithic brows before, but this particular evening followed an entire day spent reading about Neanderthals, that extinct race of human relatives whose place in our evolutionary past is the subject of this chapter.

Neanderthals are the stuff of legend. A pop-culture icon, they are usually depicted—probably quite unfairly—as primitive and brutish cavemen, semi-naked, hairy, with blank expressions, able to communicate with little more than grunts and gestures. Certainly, the name has on several occasions made a convenient insult to throw at members of my family when they display some of their more repulsive behaviours.

But Neanderthals are icons in the scientific world too—for a different reason. The discovery of the first Neanderthal skeleton in 1856 sparked one of the longest-running and most heated debates in modern science, between those who believe the Neanderthals are human ancestors, and those who are adamant that they are merely an extinct side branch of the human evolutionary tree.

Despite many decades of effort, scientists had almost given up hope that the place of Neanderthals in human evolution would ever be known for sure. But then a remarkable piece of ancient DNA research was carried out which would revolutionise the way we view both Neanderthals and our own species.

The reason that my research on Neanderthals led me to notice Andrew's eyebrows was obvious, as there is one particular feature that Neanderthals are famous for—their huge brow ridges. Combine my day's reading with the poor light in our lounge room casting shadows on everything, and Andrew's bony brows didn't stand a chance of remaining unnoticed. The trouble is, now I can't stop staring at them!

Discovery of the Neanderthal

It was a peaceful late summer day in Germany's Neander Valley in 1856. In the valley was a limestone quarry. For the workers there, the day began as ordinarily as any other. The team had set itself the task of extracting material from two grottoes high on a cliff above the Düssel River, which meanders across the valley floor below. The grottoes were very difficult to reach, both from above and below, and so were the last in the valley to be touched.

The workers began to excavate, but they did not get far before they uncovered a number of bones. There was a skullcap, some thigh bones, ribs, arm and shoulder bones, and part of a pelvis. The workers thought perhaps they were the remains of a cave bear.

Thinking that the bones might be of some interest, they decided to pass them on to the local expert, Carl Fuhlrott. Fuhlrott was a mathematics teacher at the nearby school, but he was also well known in the area for his interest in natural history and his collection of 'curiosities'.

Fuhlrott was struck by the thick, bowed leg bones and the protruding brow ridges on the skullcap, and immediately realised that these were not cave bear bones. He had the radical thought that they might instead belong to a primitive type of human. This was a daring hypothesis, considering the prevailing view of human origins in Europe at the time.

In 1856, when the skeleton was uncovered, European beliefs about human origins stemmed almost entirely from theological tradition. This was in keeping with most other cultures on Earth, which traditionally have a view on human origins that involves a god or higher being. The Christian church taught that humans were created by God in his own image, just a few thousand years ago. Humans were placed by God at the head of a 'great chain of being', and were followed in this chain by all other 'lower' forms of life on Earth, including animals, plants and insects. Humans were the only creatures believed to have a soul.

Most people at the time did not believe in evolution, the process by which a species or race gradually changes over time into one or more different forms. In fact, church teachings specifically stated that species did not change over time, that they had been exactly the same since they were first created. Most people certainly did not think evolution applied to humans—the very idea was preposterous.

Indeed, there was no compelling reason at the time to believe anything other than church doctrine. Before the unusual skeleton was discovered in the Neander Valley, although a few suspicious-looking fossils had been found, no human ancestor species had actually been recognised. Thus there was no real evidence that humans had ever been anything other than what they are like today, nor was there a solid, scientifically rigorous theory of evolution. Incredibly, the skeleton was uncovered just three years before Charles Darwin published his revolutionary theory of natural selection, which in itself would completely change the way many people thought about the origins of all species on Earth, including humans.

That is not to say that no one had ever thought about evolution before Darwin. Various evolutionary theories had been proposed throughout history, the earliest recorded dating back to the ancient Greeks, some 2500 years ago. However, most of these early evolutionary thoughts were only partially 'scientific', and were interwoven with mythical and religious ideas.

Well before the nineteenth century, science as we know it today began to develop. But evolutionary ideas did not feature in the new discipline of 'natural history', the study of the natural world. Early science was intimately intertwined with religion, and the study of nature was conducted primarily with the purpose of learning more about God's intricate design for the species he had created.

By 1856, the strict religious view was being challenged by some in the scientific world, and the first inklings appeared that species might have changed over time. Evidence for this was the array of fossils being discovered, including enormous dinosaurs and other previously unknown forms of life.

In the years before Darwin published his revolutionary theory, a number of alternative evolutionary theories were suggested. However, each of these had its problems. Some could not be tested scientifically, and others were shown to be false when subjected to scientific testing. Most theories were supported by only a limited number of specialists. The reality was that when the Neander Valley skeleton was discovered, some members of the scientific community were beginning to take the possibility of evolution seriously, but the vast majority of people still strongly followed the religious teachings of creation. Even those who did suspect that some species had changed over time certainly did not believe that humans were in this category.

So when Fuhlrott suggested that a skeleton with such obvious differences to modern-day humans might be some type of human ancestor, and therefore that humans might have changed over time, he showed what a revolutionary thinker he was. Fuhlrott was so intrigued by his idea that he called in Hermann Schaafhausen, a professor of anatomy at the nearby University of Bonn, to give his opinion on the bones. Schaafhausen agreed with Fuhlrott's suggestion that the skeleton belonged to a primitive type of human. Schaafhausen thought perhaps it came from a race of people ancestral to the Celts and Germans.

Together, Schaafhausen and Fuhlrott presented their hypothesis at a gathering of the Lower Rhine Medical and Natural History Society in Bonn, and with that the great Neanderthal debate began.

The great Neanderthal debate

In light of the accepted view of human origins at the time, it is not surprising that the skeleton, with its unusual features, and Schaafhausen and Fuhlrott's assessment of it, caused quite a stir.

The skeleton burst into a scientific world already divided over evolutionary matters, and at first only helped to emphasise the differences of opinion. There was no reliable method available at the time to determine how old the bones actually were, so it was really anyone's guess what the skeleton was.

On the one hand, those who already had leanings towards evolutionary ideas believed that the skeleton was of great antiquity, and eagerly agreed that what had been found was the first evidence that humans had changed through time—in other words, evolved—just as they believed other species had done. Others were not so convinced. They thought that the skeleton was fairly recent in origin, and had belonged to an individual who had suffered from a variety of physical deformities throughout his or her lifetime, perhaps due to a severe disease such as rickets.

One of the more amusing suggestions put forward at the time was that the bones had belonged to a Russian Cossack, who had died when sheltering in the cave from an approaching army. He had bowed legs because he had spent years riding horses, and his huge brow ridges were due to his constantly furrowed brow, his reaction to intense stress. How exactly he had come to shelter in a cave some 18 metres up a cliff face, stark naked and without his weaponry, was never explained.

Hoping to end the argument once and for all, Rudolf Virchow, a German anatomist and pathologist, conducted a thorough analysis of the bones. Virchow was a prominent figure in German science, but he happened to have an intense dislike of the idea of evolution. It is perhaps not surprising that he concluded that the skeleton was not ancient after all. Instead, he professed, the appearance of the bones was indeed due to some form of illness. He suggested, like others before him, that the person the skeleton belonged to had suffered from rickets as a child, which explained the bowed legs. The large brow ridges, he concluded, were caused by repeated blows to the head.

As a result of Virchow's analysis, most experts at the time initially came to agree that the enticing view into our evolutionary past offered by the skeleton was imaginary—the skeleton was nothing more than that of a fairly recently deceased unfortunate, who must have endured a life of great pain and suffering.

This might explain one odd-looking skeleton, but it wasn't long before fossil enthusiasts began to uncover skeletons with the same unusual characteristics from a whole range of sites scattered throughout Europe and Western Asia—and to reassess a couple of earlier discoveries. Some were found with stone tools and the remains of extinct animals, the first hard evidence that the skeletons themselves might indeed be ancient.

The idea that the large brow ridges and heavy, bowed limbs, now seen time and time again, could be due to disease began to seem extremely unlikely. Whoever they were, it became obvious that these strange beings had been a significant population at some time in the history of that region of the world. They were even given a name, 'Neanderthals', after the Neander Valley where the first identified skeleton was found.

In 1859 the Neanderthal cause was strengthened by the publication of Darwin's famous book, *The Origin of Species*, in which he outlined his groundbreaking evolutionary theory of natural selection. Perhaps unsurprisingly, Darwin's 'heretical' ideas immediately met with fierce opposition, especially amongst the religious community. However, he also had some influential supporters. The result was a series of fierce debates, and evolution became a very popular topic for discussion in society. The ultimate outcome was that, although the idea of evolution remained anathema to many, a much larger percentage of the population began to accept the possibility that not only plants and animals, but humans too, evolved.

Nineteenth-century Neanderthal experts still thought of their subjects as primitive and brutish cavemen, quite inferior in all aspects to modern-day humans. But in the light of overwhelming evidence, they were soon forced to agree that Neanderthals were indeed a significant race of people who had some place in human prehistory. Many questions remained unanswered, however. Where did they come from? Why did they suddenly disappear? Could they have been the direct ancestors of modern-day Europeans? Or were they simply a side-branch in the human evolutionary tree, a race doomed to extinction?

Intriguing questions indeed, and a large number of people became determined to find answers. As the realisation began to dawn that humans really *had* evolved, hunting for fossilised human ancestors became extremely fashionable, and a virtual digging frenzy ensued. The secrets of humanity's past, hidden for so long inside the Earth, were about to be revealed.

More discoveries: Homo erectus

Fossicking for fossils began in earnest and more Neanderthal remains were soon uncovered, adding to the growing wealth of knowledge about this intriguing race. But even more interestingly, fossil hunters began to find remains from a range of other previously unknown human ancestors. It soon became obvious that humans as a species had a rich and complex family tree. All over the world, the scattered pieces in the jigsaw of human evolution began to emerge.

The very next human ancestor to be discovered turned out to be an important player in the Neanderthal drama. Eugene Dubois, a Dutch scientist and evolutionary enthusiast, began an expedition in the 1890s to search for human ancestors in Indonesia. His rationale for searching in this part of the world was that chimpanzee and orang-utan fossils had recently been found there. As these apes are the closest living relatives to humans, Dubois thought, logically, human ancestor fossils must also be located there.

As it happens, he was correct. After excavating at several sites with no success, Dubois and his team finally uncovered some quite remarkable fossils on the banks of the Solo River in Java: skeletal remains that were similar to Neanderthals, with the same brow ridges and robust skeletal features, but with a noticeably smaller brain. An even older human relative had been found. The new ancestor was named *Homo erectus*, or 'Upright Man', and more fossils, similar to the ones in Java, were soon found in many other regions of the world.

Apart from being an interesting human ancestor in its own right, *Homo erectus* is particularly important to this story because it turned out to be the key to the *origin* of the Neanderthals. Based on the fossil evidence, it is believed that *Homo erectus* evolved in Africa from another even older human ancestor, at least 1.5 million years ago. Just like modern-day humans, *Homo erectus* became a very effective coloniser and eventually moved out of Africa, to spread throughout Asia, the Middle East, India and Europe. As it turned out, *Homo erectus* was the earliest human relative ever to travel about the world. All previous human ancestors—and there were quite a few before *Homo erectus* appeared—had been confined to the African continent.

As time went on, researchers realised that the *Homo erectus* remains found in different parts of the world exhibited slightly different skeletal features. These were much like the small variations that can be seen in people all over the world today, but it was even more pronounced in the various races of *Homo erectus.* These 'racial' differences make some researchers think that the races of *Homo erectus* were so different that they could have even been separate species, unlike modern-day humans, who are all members of the one species, *Homo sapiens.*

And here is the key to the origin of the Neanderthals. The fossils that have been found clearly show that they evolved around 300 000 years ago as descendants of one of the races of *Homo erectus* that lived in Europe and the Middle East. Exactly which race gave rise to the Neanderthals is still debated, but essentially the discovery of *Homo erectus* solved one part of the Neanderthal mystery—the mystery of where they came from.

Neanderthal lifestyle and culture

Along with this insight into their origins, a detailed picture also began to emerge about the Neanderthal lifestyle: when and where they lived, as well as a glimpse of what their personalities might have been like.

We now know that the Neanderthals were a race of prehistoric people who lived in an area stretching right across Europe and Western Asia as well as in parts of the Middle East. They first appeared from a race descended from *Homo erectus* as long as 300 000 years ago. For thousands of years they lived in this part of the world, but around 30 000 years ago they disappeared forever.

Neanderthals are usually portrayed as brutish, primitive, rude and vulgar, with huge muscles, terrible manners and very low intelligence. They are often shown in a stooped posture, almost as if they are too stupid to drag themselves into a fully upright position. This is certainly what the first people to discover them thought, and this perception of the Neanderthals has persisted, helped along by cartoons, books and general opinion. But were they really like this?

It is true that Neanderthals were extremely strong: although they were somewhat shorter than humans, they were stocky, with large muscles. In particular, they had huge jaw muscles, and must have had an enormous bite strength. In fact, the major way in which Neanderthals differed in appearance from modern humans was in the skull. Neanderthals had large faces, low foreheads and huge brow ridges. If you feel your own chin, you will find a bony piece poking out at the bottom of your jaw. Neanderthals did not have this, which is why their chins appeared to recede.

It is not true, however, that they had a stooped posture. That particular myth is the result of an unfortunate coincidence. It just so happened that the first Neanderthal skeleton to be examined in detail was, although researchers didn't know it at the time, from an individual who had suffered a debilitating disease during his or her lifetime which resulted in a stooped appearance. With no basis for comparison, researchers assumed that all Neanderthals were

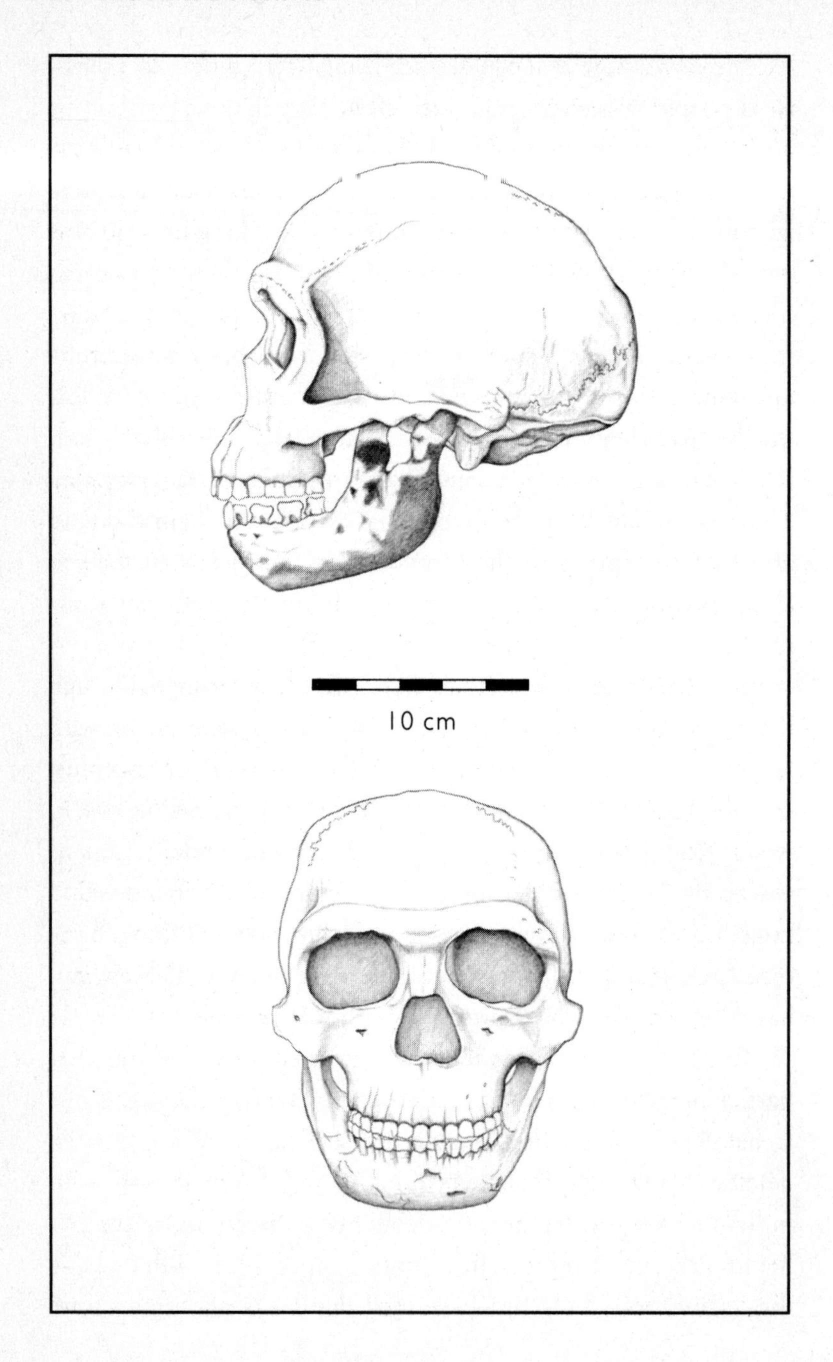
10 cm

like that, and even when more skeletons which were clearly not stooped were found, the myth persisted.

There is also no real evidence that Neanderthals were mentally inferior to modern-day humans. In fact, Neanderthal brains were slightly larger than ours. There is no way of knowing for sure how Neanderthal thought processes worked, but the assumption that they were 'a couple of sandwiches short of a picnic' probably reflected the fact that nineteenth-century Europeans generally believed *anyone* different from them must be inferior. Again, this view of Neanderthals has persisted.

Neanderthal tools were relatively simple, and Neanderthals did not make intricate ornaments, cave paintings or jewellery, as prehistoric humans did. This has always been taken as further evidence of the inferior Neanderthal brain. But it was recently discovered that the earliest humans, with brains every bit as developed as ours, did none of these things either. It wasn't until humans had been around for quite a while that these aspects of culture started to develop. The Neanderthals' simple tools and lack of art, once considered a sign of inferior brain power, can therefore no longer be thought of in this way.

Neanderthals had one other trait which is usually considered to be uniquely human. There is evidence that they buried their dead, something no animal apart from humans is known to have done. This indicates that Neanderthals might actually have cared deeply for one another, which is far from the traditional view of how they might have behaved.

Opposite: *A Neanderthal skull, showing the characteristic brow ridges and lack of a chin. Compared with a modern human, the skull is longer from back to front, and rounder at the sides. The average size of the Neanderthal brain was slightly larger than that of a human.*

What became of the Neanderthals?

Unravelling the mystery of the origin of the Neanderthals, and learning more about their lifestyle, personality and distribution, proved to be interesting and fruitful areas of scientific research. However reaching a consensus as to what *became* of the Neanderthals turned out to be much more difficult and controversial. Were they indeed the direct ancestors of modern Europeans, as Fuhlrott and Schaafhausen had first suggested? Did they therefore disappear from the fossil record simply because they evolved into humans? Or were they distant cousins who became extinct without leaving any descendants?

It soon became apparent that these seemingly simple questions would be extremely problematic to answer; from the beginning, ideas about the fate of the Neanderthals were continually tossed back and forth on a tide of controversy. The fate of the Neanderthals turned into one of the longest-running and most heated scientific debates in history.

As time passed, a variety of fads came and went: one minute Neanderthals were hailed as the ancestors of humanity and the next they were shoved aside as merely an extinct cousin, a withered branch to nowhere on the family tree. Try as they might, researchers could not come to an agreement on the fate of the Neanderthals, or on their relationship to modern-day humans. Over the decades that followed, new discoveries and new methods of analysis were developed, but often these only added to the debate. The trouble was that the available evidence was just too ambiguous.

By the mid-1980s, after almost 150 years of intense research and debate, the issue of the Neanderthals' fate was no closer to resolution. Two fiercely opposing sides had formed in the argument, one believing the Neanderthals to be human ancestors, the other adamant they were not.

More arguments over fossils

By this stage, researchers had at their disposal not only an abundance of fossils, but also a variety of innovative new dating methods—tools that would have been unimaginable to the earliest Neanderthal enthusiasts. Despite this, the two groups of researchers were as directly and fiercely opposed to each other's theories as earlier generations had been, for both sides could see plenty of evidence in the fossil record to support *their* side of the argument.

At the forefront of one side of the modern-day Neanderthal debate was Chris Stringer, a palaeontologist from the Natural History Museum in London. Stringer and his supporters strongly believed that the fossil evidence indicated that Neanderthals could not possibly have been direct human ancestors.

The group proposed a scheme for the fate of Neanderthals which became known as the 'Out of Africa' hypothesis. The hypothesis began with the time just before modern humans evolved, when Earth was populated by the various races descended from *Homo erectus.* Neanderthals, of course, were one of these races. According to Stringer's scheme, each race had evolved to the point where it was distinctly different from all the others—so different, in fact, that each was actually a separate species.

Stringer and his colleagues proposed that modern humans then evolved *in Africa only*, springing as a new species from just one descendant species of *Homo erectus* that had existed there. Over the tens of thousands of years that followed, modern humans then spread throughout the rest of the world, replacing all the other descendants of *Homo erectus*, including Neanderthals. Because they were separate species, no interbreeding could occur between the invading humans and the other *Homo erectus* descendants, meaning that Neanderthals, and any other descendants of *Homo erectus* outside Africa, are in no way the ancestors of humans.

How the proposed takeover might have happened is not entirely clear. However, it is possible that the earlier humans might have fought and killed all the races of *Homo erectus*. Alternatively, they might simply have out-competed them for food and resources. Unfortunately, this theory has a dismal end for the Neanderthals—they simply became extinct.

Stringer and his colleagues believed that the fossil record supported their theory entirely, showing an abrupt difference between the skeletons of the various races of *Homo erectus* and the skeleton of a modern human, thereby proving that it was not possible for *Homo erectus* to have evolved into modern humans in most regions in the world. The one exception to this, Stringer said, was in Africa, where the *Homo erectus* fossils did show a smooth transition to modern humans, in perfect agreement with their theory.

In stark opposition to Stringer and his colleagues was the group led by Milford Wolpoff, a palaeoanthropologist from the University of Michigan. Wolpoff and his colleagues believed quite strongly that Neanderthals *were* one of humanity's ancestors—and that they had evidence to prove it.

Wolpoff's theory also began just before humans evolved, when Earth was populated with the various races descended from *Homo erectus*, including Neanderthals. However, in complete contrast to Stringer, Wolpoff and his colleagues believed that all the different races of *Homo erectus*, although subtly different from each other, *were* still members of the same species, and were therefore able to interbreed freely.

Over many thousands of years, or so this theory goes, the different races of *Homo erectus* continued to evolve, changing ever so slightly, but never losing their ability to interbreed. Eventually, thousands of years ago, they became what we would consider to be human. Humans did not evolve only in Africa, from one *Homo erectus* descendant, said Wolpoff, but from a mixture of *Homo erectus* descendants all over the world.

According to this model, Neanderthals have an important place in human evolution as one of humanity's ancestors, together with a number of other races of *Homo erectus.* Wolpoff and his supporters believed that each human race is descended primarily from a slightly different race of *Homo erectus*, and this is the cause of the racial differences that can be seen amongst humans today. This would mean the differences that exist between the various human races are very old, dating back a million years or more to when *Homo erectus* first began to split into different races. For obvious reasons, this theory became known as the 'Multiregional Evolution' hypothesis.

Like Stringer, Wolpoff and his colleagues found support for their theory in the fossil record. It was all a matter of how the fossils were interpreted. Wolpoff did not agree with Stringer's belief that *Homo erectus* fossils outside Africa were starkly different from modern human fossils. He believed that Neanderthal and other human ancestor fossils over the past 1.5 million years or so showed a gradual transition from *Homo erectus* to modern humans all over the world. For example, he believed that modern-day Asians closely resemble the *Homo erectus* fossils found in that region, and that modern-day Europeans resemble Neanderthal fossils. From Wolpoff and his colleagues' point of view, the fossils confirmed that Neanderthals, along with the other *Homo erectus* races, can be considered the ancestors of humans.

Based on fossil evidence alone, an impasse had again been reached in the debate about the fate of the Neanderthals, and was proving very hard to resolve. After almost 150 years of debate, and still no agreement, it seemed that the issue would never—in fact, *could* never—be solved.

Human DNA extraction and its implications

In 1987, the issue of the fate of the Neanderthals and their relationship to humans took another rather exciting twist when some completely new evidence was supplied—evidence which, for the first time, did not rely on the troublesome human fossil record. In a remarkable announcement in the prestigious scientific journal *Nature*, Berkeley researchers Rebecca Cann, Mark Stoneking and Allan Wilson presented an innovative and clever method to investigate the question of human evolution in a completely new way: by looking for clues in DNA extracted directly from people alive today.

Carefully and painstakingly, the researchers had collected tissue samples from 147 modern-day people who originated from a range of different geographical locations, including Africa, Asia, Australia, Europe and Papua New Guinea. Using the same basic process described in the Introduction, the scientists extracted DNA from each sample, then compared specific sections of the DNA. What they found was the startling fact that all humans, whether from Asia, Africa or Europe, or anywhere else for that matter, have astonishingly *similar* DNA. This simple finding may seem rather trivial on the surface, but in fact it had great implications not only for the debate about the fate of Neanderthals, but for the origins of humanity itself.

To make sense of what Cann, Stoneking and Wilson's research indicated, I like to think of a story that my mother often tells about the day I was born. Having first established that I was a girl and that I was well and truly alive, my mother 'gave me the once over' to see what I looked like. As a result of this understandable curiosity, one of the first sentences to fall on my ears was: 'Oh my God! She has her grandpa's toes!'

The point of this story? That from the moment we are born, it is obvious just how closely each of us resembles our nearest relatives. The same is also true of DNA, the genetic material inside each and every cell which determines many aspects of what we look like and how we behave—in essence, who we are. DNA is passed on from parents to offspring and ancestor to descendant but, over time, mutations, or changes, tend to occur in the DNA. The result is that the more closely related two living things are to each other, the more similar their DNA is likely to be. It follows, therefore, that if two living things have very similar DNA, it is a pretty sure bet that they are close relatives, and shared a common ancestor quite recently.

Cann, Stoneking and Wilson's research showed that all humans have remarkably similar DNA—the section they looked at differed by an average of only about 0.5 per cent amongst all the 147 people they sampled. This implied that modern-day humans are all very closely related, and are therefore likely to have had a very recent common ancestor.

As if this result alone were not enough, Cann, Stoneking and Wilson then performed a clever piece of mathematical analysis to calculate when and where this common ancestor of all humans is likely to have lived. The method they used is actually quite straightforward: previous analysis of human and animal DNA had indicated that, due to the naturally occurring mutations that are a feature of the DNA of all living things, the section of DNA they had focused on tends to evolve (change) at a rate of 2–4 per cent per million years. The fact that all the DNA in their samples differed by an average of about 0.5 per cent indicated that the common ancestor of all humans probably lived somewhere in the region of 140 000–290 000 years ago. Furthermore, because of the patterns of similarities and differences between the DNA in their samples, the researchers were able to postulate that this ancestor lived in Africa.

On the face of it, this DNA evidence seemed squarely to defeat Wolpoff and his supporters' view. Their theory stated that humans had arisen in multiple regions of the world, from multiple *Homo erectus* descendants, one of which was the Neanderthals. This implied that the common ancestor of humans was much older—dating back to early *Homo erectus* times, which could be as long ago as 2 million years or so. If this were the case, DNA from modern-day humans should be much more diverse, as there would have been much more time for mutations to occur.

The view of Stringer and his colleagues, on the other hand, matched perfectly with the results of Cann, Stoneking and Wilson's DNA work. The Out of Africa hypothesis said that humans evolved recently in Africa only, and were descended from just one, non-Neanderthal race of *Homo erectus*—exactly what Cann, Stoneking and Wilson's results indicated.

Did this mean that the Neanderthal debate was over at last? Unfortunately, not quite. Wolpoff would not accept the results of the DNA work for a moment, and certainly was not going to be silenced easily. Stating that 'there isn't a snowball's chance in hell' that Cann, Stoneking and Wilson were correct, he insisted that the fossils do not lie—that they clearly show a transition from *Homo erectus* to humans everywhere in the world that one would care to look—and that no modern human DNA would change this. 'If you really want to know where modern humans come from, go look at some fossils,' he said defiantly.

As well as his unwavering faith in what the fossils seemed to be telling him, Wolpoff also questioned some of the methods of analysis the researchers had used in their study, and reiterated his belief that the results did not accurately reflect what really happened when humans evolved. His supporters agreed with him. So the debate continued.

Neanderthal DNA enters the scene

An impasse had again been reached. The difficulty was that comparing DNA from modern-day humans, although innovative, was an indirect method of looking at the past. There would always be questions of interpretation, and debate over whether modern-day DNA really can accurately reflect past events.

The DNA evidence did seem to support the Out of Africa hypothesis that Neanderthals could not have been human ancestors, but it was clear that some more direct evidence was needed if the debate were to be solved once and for all.

It wasn't long before researchers took the next logical step. What if it were possible to extract DNA directly from a Neanderthal bone, and compare it to DNA from modern-day humans? If Neanderthals really were the direct ancestors of humans, in particular those people still residing in the areas where Neanderthals had lived (which included Europe and Western Asia), they would have passed their DNA directly on to their modern human descendants, and it should be quite similar to the DNA of modern-day Europeans and Western Asians.

On the other hand, if Neanderthals were *not* the direct ancestors of Europeans, but only a more distant relative, Neanderthal DNA should be quite different to that of modern Europeans. If Neanderthal DNA could be extracted and analysed, perhaps the debate could finally be resolved.

No one could have been in any doubt that the task of extracting and analysing Neanderthal DNA would be one of the most challenging ancient DNA projects ever attempted. So Neanderthal researchers decided to approach one of the pioneers of ancient DNA research, Svante Pääbo, from the Zoological Institute at the University of Munich, Germany.

Born in Sweden, Pääbo had studied archaeology at university level while still at high school, but found the work 'too slow' and turned to the study of immunology. Then, while a graduate student, he began to dabble in the brand-new field of ancient DNA research in his spare time. After some convincing, the director of an East Berlin museum allowed Pääbo to experiment on the museum's Egyptian mummy collection, and he began trying to extract DNA from the ancient remains at night and on weekends.

Pääbo obviously found his niche with ancient DNA, and continued to develop this emerging field of research, first at the University of California, Berkeley, and then at the Zoological Institute of the University of Munich. By the time the Neanderthal DNA work began, he had made quite a name for himself, and was the perfect person to take charge of such a difficult task.

Together with his graduate student Matthias Krings, Pääbo removed a tiny 3.5 g sample of bone from the right humerus of a Neanderthal. Fittingly, the bone was from the original skeleton found in 1856 in the Neander Valley. Painstakingly, the pair extracted the DNA from the sample, using the same basic process described previously. Next, they compared sections of the Neanderthal DNA with the equivalent sections of DNA from a number of present-day human samples, from people from different regions of the world.

A DNA strand is a long, thread-like molecule, made up of a string of smaller units, called 'bases', which are linked together to form the strand. There are four different types of bases: adenine (written as A), thymine (T), guanine (G) and cytosine (C). In the cells of organisms, two strands of DNA are wound together to form the famous 'double helix', which is in turn wound and bundled further to form the shape of the chromosomes (see diagram, p. 3).

In order to compare the DNA in two different organisms, as I mentioned earlier, DNA is first extracted from samples of each, and copies are made in the laboratory of the particular segment of DNA from each sample that researchers are interested in. The next step is to determine the DNA sequence—the order of bases in the DNA strand—from that particular sample. This is done in a process that uses a variety of chemicals and enzymes, and which these days is often performed in an automated fashion by large machines known as DNA sequencers. Once the DNA sequences of each sample are known, they can be compared and analysed. At this stage, it is more or less a matter of simply lining up the sequences on a computer screen and looking for similarities and differences.

When Pääbo and Krings lined up their sequence of Neanderthal DNA with the sequence from the equivalent section of modern human DNA, there was no mistaking the results. Neanderthal DNA and human DNA were quite different. To be exact, Pääbo and Krings found the Neanderthal DNA varied from human DNA sequences by an average of 26 individual differences. To be sure the results were correct, researchers Anne Stone and Mark Stoneking repeated each step of the procedure at an independent laboratory at the Department of Anthropology of Pennsylvania State University. The DNA they extracted was exactly the same, which confirmed Pääbo and Krings' results.

GGAAGCAGTTGGGTACCACCCAAGTATTGACTCACCCATCAACAACCGCTATGTATTTCGTACATTACTGCCAGCCACCATGAATATTGTACGGTA

GGGAGCAGTTGGGTACCACCCAAGTATTGACTCACCCATCAGCAACCGCTATGTATCTCGTACATTACTGTTAGTTACCATGAATATTGTACGATA

*Section of human (left) and Neanderthal (right) DNA sequences. Differences are shown in bold in the Neanderthal sequence. Only one strand from each DNA double helix is shown—the other can be inferred using the base pairing rule (*A *pairs with* T*, and* C *pairs with* G*). (This diagram is adapted from the sequence shown in Krings, M. et al (1997), 'Neanderthal DNA sequences and the origin of modern humans',* Cell, *vol. 90, pp. 19–30)*

A conclusion to the Neanderthal debate at last?

There was only one conclusion that Pääbo and his team could make. If Neanderthals truly were human ancestors, their DNA would have been much more similar to modern human DNA. Instead, their work provided the first truly clear evidence that Neanderthals could not possibly have been human ancestors.

As soon as the results of the Neanderthal DNA work were announced, debate erupted again with a vengeance. Everyone had something to say about what they thought of the results, and what it meant for the Neanderthals. Was this finally absolute proof that Stringer and his colleagues were right, and that Wolpoff and his colleagues were wrong? Was the great Neanderthal debate finally over?

Stringer and his colleagues were understandably delighted with the results of the experiments. Stringer called the DNA work 'a terrific achievement', one which, in his opinion, provided compelling evidence that he and his supporters had been correct.

Not so fast, said Wolpoff and his supporters. Although agreeing that the Neanderthal DNA was 'an extremely important piece of work', Wolpoff pointed out that there was only one sample of Neanderthal DNA so far, and more would be needed before any definite conclusions could be drawn. Some other aspects of the analysis of the Neanderthal DNA bothered him too. 'It's not that I want to rain on anybody's parade,' he said, 'but there are some nagging details.'

Although no one could deny that a truly revolutionary piece of work had been carried out, it became apparent that a single Neanderthal DNA sample was not going to end the debate. More Neanderthal DNA, from a different individual, needed to be extracted and compared with Pääbo's results.

Thus it was that a short time later, a second team of scientists, led by researcher Igor Ovchinnikov, extracted DNA from a second Neanderthal specimen, a child found in a cave in southern Russia, one of the easternmost Neanderthal populations. Despite the geographic separation between the samples, when the DNA of the Russian Neanderthal was compared with the DNA from the first Neanderthal, it proved to be very similar. Like the DNA from the first Neanderthal, the Russian Neanderthal DNA was also very different to the DNA of modern humans.

Krings, Pääbo and their colleagues also extracted DNA from Neanderthal bones found in a cave in Croatia. Again, the DNA was similar to the previous two samples, and quite different to that of modern humans.

Three Neanderthal DNA samples now showed that Neanderthal DNA was significantly different to modern human DNA. Despite this evidence, Wolpoff and other Multiregional Evolution hypothesis supporters still did not accept that it had been conclusively proven that Neanderthals were not human ancestors. They pointed out a fundamental flaw in the research—that Neanderthal DNA was being compared with DNA from *present-day* humans, not DNA from humans living several thousand years ago, closer to the time that the Neanderthals had disappeared. What if, Wolpoff suggested, human DNA has changed since that time? If this was the case, then maybe DNA from early true humans would be much more similar to Neanderthal DNA, meaning they could be human ancestors after all.

Wolpoff had a good point, researchers on both sides of the debate agreed—until there was a good picture of not only what Neanderthal DNA was like, but also the DNA from the earliest true humans, we wouldn't know for certain. Extracting and comparing ancient human DNA and Neanderthal DNA would help to clear up any nagging doubts created by comparing modern human DNA with ancient Neanderthal DNA.

This led to a group of Italian and Spanish scientists extracting DNA from a sample of bone from a 25 000-year-old fully modern European human from the Paglicci cave in southern Italy. Pieces of DNA from this skeleton were compared with DNA from present-day people and to Neanderthal DNA. The DNA showed just what Stringer and his supporters had suspected: the DNA from the ancient modern human matched present-day human DNA but did not match Neanderthal DNA. This, surely, makes it extremely unlikely that the Neanderthals were human ancestors.

Conclusion: Neanderthals are not human ancestors

In a stunning example of the power of ancient DNA research to provide answers to fascinating real-life issues, the Neanderthal DNA work has finally made it possible, after 150 years of debate, to say with some certainty that the enigmatic Neanderthals are not the ancestors of humans, but are simply an example of an extinct species, albeit an extremely interesting one.

As we are all too aware, species extinction has been an ongoing natural phenomenon throughout the evolutionary history of life on Earth. Some estimates suggest that as many as 99 per cent of all species that have ever lived are now extinct.

A T C G

Although species become extinct, they do not always vanish without a trace. Sometimes, by lucky coincidence, when an organism dies its remains become preserved. Because of this, it has been possible to find the bones, teeth, fossils and sometimes entire preserved carcasses of extinct species.

Some of these remains still contain DNA. Research on the DNA from extinct species is in fact one of the most productive areas of ancient DNA work and, just as in the case of the Neanderthals, is regularly used to investigate the relationships between extinct species and their living relatives.

While this is an interesting and productive area of research in itself, it has paved the way for exploration of an even more intriguing proposition: whether DNA could be used to bring an extinct species back to life. Might we one day create a 'prehistoric zoo' in which monkeys mix with mammoths, and tigers with thylacines? Does extinction have to mean forever?

2

THE PREHISTORIC ZOO

Could extinct animals be brought back to life?

I was brought up in New Zealand, and every school holidays when I was a child I would stay with my grandparents in Wellington. Nanna would spoil me, buying me treats and making my favourite meals, and Grandpa would take me to all my favourite places. We would visit the botanic gardens, where I would feed bread to the ducks and then race up the hill to the adventure playground. We would go for a ride in the cable car, an icon of the city of Wellington, which had polished wood panelling and handles hanging from the ceiling that I could never reach. We would go to the zoo, where I would giggle as I watched the monkeys search each other's backs for fleas. But the highlight of every stay was definitely our trip to the museum.

The museum was a magical place. There were all manner of things fascinating to a curious 6 year old—Maori canoes, a real Egyptian mummy, even an artificial reef complete with suspended plastic fish. Around every corner there was something different and interesting to look at. But every time I walked through those halls, there was one thing I wanted to see more than anything else.

We would walk through exhibition rooms, down corridors, and up and down stairs until finally we rounded a corner and there it was: the moa. This huge, extinct bird towered far above my head, above everyone's head. It fascinated and frightened me at the same time. I longed to reach out and touch it, to feel its soft feathers and leathery feet. It looked so real, as if it could leap off its platform and chase me across the room. Of course, I knew that this would never happen. Even as a little girl, I knew that extinct creatures never come back to life. After all, extinction is forever, isn't it?

The remains of thousands of extinct animals, birds, insects and reptiles are housed in museums and other collections around the world. Teeth, bones, shells and, in a few lucky cases, entire preserved carcasses have been found, buried in caves or deep underground. These relics from the past make up the exhibits in these huge prehistoric zoo collections. But unlike zoos filled with living animals, the rooms that house these relics are eerily silent and still. There are no chirps, squeaks, roars or growls, no flying, hopping or swimming. The animals in these collections, once vibrant and alive, are no more.

The array of different species that once roamed the Earth is simply breathtaking. This is not surprising when you consider that the total number of species that have become extinct since life first evolved far outweighs the number of species in existence today. Some, such as the Australian thylacine, for example, have been extinct only a short time. The last thylacine died an unfortunate and untimely death in Hobart Zoo in the 1930s, meaning there may still be some people alive today who are lucky enough to remember seeing one.

Other species, such as the woolly mammoth, have been extinct for many thousands of years. Some species have been extinct for millions of years, most notably the dinosaurs. But it doesn't stop there. Throughout the entire 3 billion years or more that life has existed on Earth, as new species have evolved others have in turn become extinct.

Considering that most extinct species have never been seen alive by anyone around today, it is remarkable what intricate details have been gleaned about them. Detailed analysis of bones, shells, teeth and carcasses has allowed palaeontologists to piece together details of the animals and their lives: what they ate, what their habits were, what they looked like, when they first appeared, and when and why they became extinct.

Traditional analysis involved examining the physical structures of the remains, from which a wealth of information could be mined. Physical analysis is still an invaluable method for learning more about the history of life on Earth, but by the 1980s a new technology was on the horizon, one which would compel those dusty remains to reveal even more of their closely kept secrets.

In recent years, huge advances in DNA technology in general had already been made, meaning that the analysis of DNA was becoming ever more sophisticated. It had become an important tool for studying various aspects of *living* species, and it was now a routine task to take a sample of blood from a living animal, extract the DNA and carry out experiments on it—for example, to examine how similar the DNA from two species was in order to learn more about their evolutionary relationships, or to investigate the DNA of individual genes from an animal to learn more about how different genes function.

This was about ten years before the groundbreaking Neanderthal DNA work discussed in the previous chapter had been carried out, and DNA had not yet been extracted from an extinct animal. Although the idea was intriguing, no one knew if DNA would even have survived in the remains of an extinct animal, let alone whether it could be extracted. The idea, however, was irresistible, and it was inevitable that someone would want to try to find out.

The quagga: Extracting DNA from extinct species

Reinhold Rau was a taxidermist at the South African Museum in Cape Town. In 1969 he began work on re-mounting a quagga foal which had died 140 years earlier. The quagga was a South African mammal which looked much like a cross between a zebra and a horse. It had zebra-like black stripes on its head and shoulders, with a plain-coloured stomach and legs. The upper part of its body was brownish, rather than the black and white colouring that distinguishes zebras.

The quagga once lived in significant numbers in South Africa, but after European settlement its numbers began to drop dramatically, for two major reasons. First, like all members of the horse family, quaggas fed on grass, which was in sparse supply. Settlers believed that quaggas were competing for the limited grazing with their own livestock, sheep and goats. Because of this, the quagga was hunted mercilessly by farmers aiming to protect their livelihood. Secondly, hunting African animals was a favourite sport in the nineteenth century.

Numbers continued to dwindle, until finally there was just one quagga mare left, kept in captivity in Amsterdam Zoo. She died on 12 August 1883 and with her the species died too. Sadly, no one realised that the quagga was extinct until many years later.

The foal Rau re-mounted had been stored in the Museum of Natural History in Mainz, West Germany. As he worked on the quagga foal, Rau found dried muscle tissue still attached to the skin. Over the years that followed, an idea began to form in his head. Could DNA be found in this dried flesh? Rau was intrigued by the possibility, and determined to know the answer.

Rau sent a sample of the muscle tissue to a group of DNA researchers in California who had expressed an interest in trying to extract DNA from it. Amazingly, the team did manage to extract some fragments of DNA, proving for the first time that DNA *could* be extracted from an extinct animal.

As a bonus, the DNA they extracted yielded some interesting information about the quagga itself. There had been a long-running disagreement as to whether the quagga was a separate, but closely related, species from all other zebras, a subspecies of the plains zebra (one of the three living species of zebra, which also includes the mountain zebra and Grevy's zebra from East Africa), or was more closely related to the horse than the zebra.

To establish which was the case, after extracting the quagga DNA the research team made copies and determined the sequence of two small pieces of the DNA, using the same basic method that the Neanderthal researchers would later use. They lined up the quagga DNA sequences with the equivalent sequences of zebra and horse DNA and compared them. What they found was that the quagga DNA was a very close match to the DNA of the plains

GGGAGGATTT TTCACTGATTCCCTCTATTCTCAGGGTACACACTCAACCAAACCTGAGCAAAATCCACTTTACAATTATATTCGTAGGAGTCAATATAACTTT

AGGAGGATT**C** TTCACTGATTCCCTCTATTCTCAGG**A**TACACACTCAACCAAACCTGAGCAAAAT**T**CACTTTACAATTATATTCGTAGG**G**GTCAA**C**ATAA**T**TTT

Section of zebra (left) and quagga (right) DNA sequences. Differences are shown in bold in the quagga sequence. Again, only one strand from each DNA double helix is shown. (This diagram is adapted from the sequence shown in Higuchi, R. et al (1984), 'DNA sequences from the quagga, an extinct member of the horse family', Nature, *vol. 312, pp. 282–4)*

zebra, with a total of only 12 base differences between the fragments from the two animals. This was close enough to prove that the quagga and the plains zebra were actually the same species, differing only enough to be considered subspecies.

This success immediately led to a rush of interest in research into DNA from extinct animal species. Would other museum specimens contain DNA in good enough condition to be extracted? How many more secrets of the evolutionary relationships of extinct animals could be revealed in this way? The hunt was on.

DNA was soon found in specimens of a range of other extinct species: mammoths, the woolly rhinoceros, pig-footed bandicoots, American mastodons, Steller's sea cows, sabre-toothed cats, cave bears, several species of sloth, thylacines, piopios, blue antelopes, New Zealand rail species, moa-nalos and lemurs. As we will see in Chapter 4, this type of ancient DNA research has given some interesting insights into the evolutionary history of my favourite extinct bird, the moa.

The Tasmanian tiger: A tale of extinction

Using the DNA from extinct animals to investigate the evolutionary relationships between them and still living species was, and continues to be, an interesting area of ancient DNA research. However, some researchers began to wonder whether something even more exciting could be done with this DNA. Could it be used to do what really seemed to belong in the realm of science fiction? Could it be used to bring extinct species back to life?

Mike Archer, director of the Australian Museum, was one such researcher. Would it be possible, Archer wondered, to bring back one particular species which became extinct quite recently: the Australian thylacine?

The thylacine is known by a variety of names, including Tasmanian tiger, Tasmanian wolf and marsupial wolf. Its scientific name is *Thylacinus cynocephalus*, which translates as 'the pouched dog with a wolf-like head'. However, the thylacine was not related to the tiger, the dog or the wolf, but was actually a marsupial.

Mammals are divided into three groups, depending on their systems of reproduction. There are the placental mammals, to which we belong; the marsupials, which have pouches; and the monotremes—the platypus and the echidna—which lay eggs. Like the kangaroo and possum, the thylacine was a marsupial. However, it was quite unlike any other marsupial alive today, and has no close living relatives.

The popular names for the animal are more suggestive of what it looked like than what it was related to. The thylacine was yellowy-brown in colour, with dark stripes across its back. In general shape it resembled a dog, and was roughly the size of a fairly large dog. An adult was about 1.5 metres from nose to tail, and weighed 30 kg.

The thylacine was once found all over the Australian continent and in New Guinea. Australia also had a range of large mammal species, known as megafauna ('large animals')—for example, the diprotodon, a marsupial the size of a rhinoceros. The megafauna provided a ready food source for predators, including the thylacine.

The first humans are thought to have arrived in Australia from Asia 50 000–60 000 years ago. At the time, the world was in the grip of an Ice Age. With much of the globe's water contained in ice, sea levels were lower, meaning that New Guinea, mainland Australia and Tasmania were one continuous land mass, making migration easier for Australia's first human inhabitants.

After humans arrived in Australia, the megafauna began to become extinct. Whether humans caused these extinctions is hotly debated; however, this is certainly one of the main theories for their disappearance.

The extinction of the megafauna drastically reduced the food supply for Australia's native predators, including the thylacine. It is possible that humans may also have hunted the thylacine for food. The thylacine disappeared from mainland Australia at least 2000 years ago, and from then on was found only in Tasmania. A range of other predator species had become extinct by then too, including several species of crocodile and a marsupial lion.

In the early 1800s, European settlers arrived in Tasmania, and began to establish farms. Rumours began to circulate that the thylacine attacked sheep and was thus a threat to the sheep farming industry, and soon the animal was being hunted in earnest. The Tasmanian Government began a bounty scheme in 1888, rewarding anyone who killed a thylacine.

The tragedy is that there is little actual evidence that thylacines ever killed more than the occasional sheep. They were certainly not the significant threat to the industry that the farmers believed them to be. However, the myth that they were sheep killers continued, and their demise was the result.

There was also a general desire at the time amongst the settlers to replace Tasmania's native plants and animals with the domestic flora and fauna they had brought with them from Europe. As the major predator amongst Tasmania's native species, the thylacine had to go.

Soon after the turn of the twentieth century, however, things started to look a little more positive for the thylacine. Attitudes were shifting, and some Tasmanians, in particular scientists and naturalists, were beginning to value the island's native flora and fauna for its own sake. The Tasmanian Field Naturalists Club was founded in 1904 and, although it supported the introduction of game species from overseas, it also promoted appreciation of native species. As part of this change in attitude, scientists began to voice concerns about the thylacine becoming extinct. In 1909, the bounty scheme was finally stopped.

But by then there were only a few thylacines left. The myth that thylacines were a threat to sheep farming had not subsided, and there was still strong opposition from farmers to the idea of saving the thylacine, hindering its conservation. Awareness of the plight of the thylacine did grow, however, and there were more calls to save the animal before it became extinct. Unfortunately, it was a classic case of too little, too late.

In 1933, Hobart Zoo bought the last known wild thylacine, a young adult female. Tragically, she died on 7 September 1936, just 59 days after the thylacine was declared a legally protected species. As far as anyone can tell, when this thylacine died, the species became extinct.

Adding insult to injury, this last known thylacine died as a direct result of human negligence. During the day she was locked out of her den, to enable the public to view her. The pen she was kept in during the day had no shelter at all, but usually she had access to a secure den area at night. In the afternoon, before leaving for the night, the keepers performed the essential task of locking the thylacine in her evening den to ensure she would be warm enough during the cold night ahead.

This was the time of the Depression, and the zoo had employed a number of unqualified workers on a type of work-for-the-dole scheme. One of these workers was assigned to be the thylacine's keeper. One winter night, the keeper neglected to lock the thylacine in her den before leaving, and she died that night from exposure. This was a regrettable period in the zoo's history, and many other animals died from neglect around the same time.

An early twentieth-century pair of thylacines in the National Zoological Park, Washington DC. (Smithsonian Institution Archives, Record Unit 95, box 49, f. 18, image # 94-12585)

Since 1936, there have been a number of reported sightings of thylacines. However, none has been confirmed, and no thylacine has been captured, nor have any remains been found. The species has almost certainly been extinct since September 1936.

Cloning the Tasmanian tiger: Can we resurrect extinct species?

The thylacine may be extinct, but not necessarily gone forever, if Mike Archer has anything to do with it. Archer, a conservation biologist, and until recently Director of the Australian Museum, has had a long-standing personal interest in Australian carnivorous marsupials, especially the thylacine.

Archer was delighted when one day he came across a preserved thylacine pup amongst the Australian Museum's collection. The pup had been stored in alcohol since its death in 1866. An idea began to formulate. Could this pup, and other thylacine remains like it, be used to clone a live thylacine?

Nothing like this had ever been done before, but Archer was determined to try. And he was not the only person to believe the task was worth attempting. Support for the idea grew and, in 1999, the Australian Museum launched a project to see if the thylacine could indeed be brought back to life. A private trust was set up to fund the research, with additional funding provided by the New South Wales Government.

The first stage of the research was to establish whether any DNA could be extracted from the preserved pup. Archer and his team had good reason to be hopeful, because of a fortuitous decision made when the pup died—it had been immediately preserved in ethyl alcohol. This was a somewhat unusual choice of preserving medium—formalin was usually used to preserve biological specimens—and was vital to the project. Formalin almost totally destroys DNA. Alcohol does not.

The project got off to a great start when the research team successfully extracted a tiny piece of DNA from the preserved pup. Following this success, they managed to extract pieces of DNA from the remains of two other thylacines.

However, the researchers were still a long way from knowing if it was possible to recreate a living thylacine. Before they could go any further, they would need to discover the minute details of not just a few small pieces of DNA, but of the thylacine's *entire genome.* In other words, they would have to uncover every detail of the DNA in each and every thylacine chromosome—an extremely complicated and expensive undertaking.

Genomes

Unravelling the details of a species' entire genome is commonly known as a genome project. The Human Genome Project is one example, and it involved millions of dollars and international collaboration between several groups of researchers. To explain it very briefly, a genome project involves several steps. First, DNA is extracted from tissue samples from a particular species, and is chopped up with enzymes into manageable pieces. Researchers then inspect each piece of DNA, bit by bit, examining every detail, to figure out where it fits in the genome. Gradually, the pieces come together like an extremely complex jigsaw puzzle, until the details of the entire genome are known.

A genome project is a very painstaking and lengthy process, even for a living species where there is a ready source of tissue samples from which to extract DNA. Only a few genome projects from living species have been completed, or are close to completion. To date, a genome project has never been successfully attempted for an extinct species. The thylacine genome project, if it succeeds, will be a world first.

If the thylacine genome project succeeds, researchers can then move on to the next step, which would be to make synthetic thylacine chromosomes containing DNA identical to that of a real thylacine. This is why the thylacine genome project is so important: without it, synthetic thylacine chromosomes cannot be accurately created, and cloning will be impossible.

In the final stage, the synthetic chromosomes would then be used to try to replace the chromosomes in an egg cell from a suitable host animal, which would act as a surrogate mother. A potential surrogate for the thylacine could be the Tasmanian devil, one of its closest living relatives.

A Tasmanian tiger by 2010?

The Australian Museum researchers hope to see the first thylacine birth by 2010, and there are signs that they are not attempting the impossible. For a start, there is a good collection of preserved remains from which DNA could, potentially, be extracted. If the genome project is to succeed, it is essential that enough thylacine DNA be found for the entire genome to be deciphered, and there is hope that this is the case. More than 60 thylacines were housed in zoos from 1856 to 1936, and there are a surprising number of preserved remains in museum collections today—all potential sources of DNA.

Another source of hope is that the thylacine has not been extinct for very long. In general, the older a specimen is, the more likely it is that its DNA will be damaged, and therefore impossible to piece together. There may be pieces missing, or the chemical structure of the DNA may be altered beyond recognition. It would certainly be easier to conduct cloning research on a recently extinct species than on one that has been extinct for a long time. The thylacine has been extinct for a mere 70 years.

There is even hope for future breeding prospects. The first preserved pup that Archer found was a male. One of the other pups found since is female.

Despite these positives, the museum admits that it might not succeed. The DNA in the remains, like the DNA in any preserved specimen, is still not in ideal condition. It still has breaks, it is fragile, it has gaps and missing pieces. Preserved DNA can never be as good as DNA extracted fresh from a living animal.

Using the Tasmanian devil as a host also poses potential problems. Although it is the closest living relative to the thylacine, there are still fundamental differences between the two species.

There is no guarantee that a Tasmanian devil would be a successful surrogate.

Cloning is difficult even in a living species. Can it really be done for an extinct one? There are many sceptics who say it is impossible. Despite this, Mike Archer is confident of success. 'Personally,' he says, 'I think this is the most exciting biological project that's going to occur in this millennium.'

Cloning and ethics

On top of the issue of whether cloning a thylacine *could* be done is the question of whether it *should* be done. For a start, the project is very expensive. Some critics say that the money would be better spent helping to save currently endangered species rather than trying to re-create one that is already extinct. There is also the problem of causing possible damage to valuable museum specimens in the effort to extract DNA. Finally, what would happen once a thylacine had been cloned? Thylacines were never bred in captivity, so even if more than one were produced via cloning, there is no guarantee that the species would be able to survive without human intervention.

Despite the criticisms, Mike Archer believes the project is definitely worthwhile: humans made the thylacine extinct, so therefore we should bring them back.

If Archer and his team are correct, in the near future it may indeed be possible to bring back animals that have been extinct for less than 100 years. But most extinct species have been so for much longer than this. Could any of them be brought back to life? Might we one day be able to go to the zoo and see a live mammoth, a woolly rhinoceros or a sabre-toothed tiger? In considering species that have been extinct for a long time as potential candidates for cloning, the mammoth may in fact be one of the best places to start.

Mammoths and the Ice Age

The mammoth belongs to a group of mammals collectively known as the Ice Age megafauna. The term 'Ice Age' refers to a geological period in the Earth's history known as the Pleistocene. The Pleistocene epoch began 1.6 million years ago, and ended about 10 000 years ago. During this time, temperatures fluctuated in a series of cycles, each lasting several thousand years. The most striking climatic feature of the Pleistocene was a series of cold, or glacial, periods which collectively encompassed most of the Pleistocene. During these glacial periods, temperatures were much colder than they are today, giving the Pleistocene its nickname, the Ice Age.

The Northern Hemisphere was the most severely affected: huge ice sheets covered large portions of northern Europe; Scandinavia was completely covered in ice, as was most of Britain; and in North America the ice covered an area as far south as New York. The situation was similar but less severe in the Southern Hemisphere: the sea around Antarctica froze, and the mountains of the Andes and New Zealand had massive glaciers.

Because so much water was locked away in the massive ice sheets and glaciers towards the North and South Poles, everything was much drier in the temperate and tropical regions of the world. The deserts were larger, there were more steppes and grasslands, and fewer forests. The distribution of vegetation was also different: many plant species were displaced from their original habitats as the ice expanded, which in turn caused a different distribution of the animals that depended on them.

Sea levels were much lower, again because much of the water in the oceans was locked into ice sheets; as a result, there was more dry land than there is today. Australia and New Guinea were connected by a land bridge that is now submerged beneath the deeper sea. A land bridge connecting Alaska and Siberia, where the Bering Strait is today, meant that animals, including humans, could move freely between the continents of Europe and America. The first human inhabitants of the Americas arrived across this land bridge during the Ice Age.

In between the glacial periods were warmer interglacial periods, with temperatures similar to today. The ice sheets reduced in size, and sea levels rose, once again submerging large areas of land. Each complete warm/cool cycle lasted, on average, 100 000 years, but the glacial periods were longer than the interglacials.

Although it is usual to talk as if the Ice Age is over, it hasn't really ended. We are now in an interglacial phase that began around 10 000 years ago. In the next few thousand years it may end, and the ice may expand once again—although some experts think global warming may have an effect on exactly when, and if, this occurs.

Aside from the climate, another outstanding characteristic of the Ice Age was the many diverse species of large mammals that lived during that time, which have become collectively known as megafauna.

There were many different species of Ice Age megafauna. In Europe and the Americas, there were woolly mammoths and cave bears, for example. Eurasia was home to the woolly rhinoceros. The legendary sabre-tooth 'tiger', with its huge upper canines, was found in the Americas, along with giant deer. In Central and South America there were giant ground sloths, some as big as elephants.

By the time the last glacial period ended 10 000 years ago, huge numbers of megafauna species had become extinct worldwide.

In Australia, the large mammals began extinction 40 000 years ago. The cave bear in Europe and the Americas soon followed, becoming extinct about 20 000 years ago. The mammoth became extinct a little later, about 11 000 years ago. And by 10 000 years ago, at the end of the Ice Age, all the giant sloths had gone. By the time the last glacial period ended, the African species of megafauna were virtually the only ones left. In total, it is estimated that by the end of the Ice Age, Eurasia had lost 28 species, and North America had lost a huge 48. Why did so many species of Ice Age megafauna become extinct?

Causes of extinction

Extinctions in general have a range of causes, and can come in a variety of forms. Sometimes numerous species are wiped out in a very short time—virtually in the blink of an eye, ecologically speaking. There have been several of these mass extinctions in the past. Perhaps the most famous is the demise of the dinosaurs, 65 million years ago. A variety of factors could have caused these mass extinctions, and often their exact cause is a matter of debate. Suggestions usually centre around a huge environmental upheaval—for example, a massive bout of volcanic activity, or the crash-landing of an enormous meteorite.

At other times, just one or a few species become extinct, or a larger number of species become extinct over a longer period of time. Causes of this type of extinction could include climate change, the spread of a new, deadly disease, or the introduction of a new predator, humans being a prime example.

So where do the Ice Age megafauna extinctions sit in this scheme? As in so many areas of science, several alternative suggestions have been made.

The original hypothesis was that many species of megafauna became extinct when the climate changed at the end of the Ice Age, and a group of researchers still supports this idea today. It is certainly true that there was drastic climate change at the end of the last glacial period, leading to vegetation change. Large expanses of grassland—the 'mammoth steppes'—became boggy areas, with grasses being replaced by moss and sedge. It is certainly possible that there was not enough food for the mammoths and other large herbivores, and that they starved to death. Evidence supporting this hypothesis lies in the fact that from 10 000 years ago until now, the world's climate has been relatively stable and relatively few mammal species have disappeared from the Americas and Eurasia during this time compared with the vast numbers that were lost during the Ice Age.

But there are researchers who suggest another hypothesis, particularly for the extinction of the American megafauna. During the Ice Age, humans arrived in the Americas for the first time and spread throughout the continents. At the same time, more than 50 species of large animals became extinct in North America. Did the arriving humans hunt them to extinction?

In North America, there is evidence that humans did indeed hunt mammoths, at least. Mammoth bones have been found bearing scrape marks which look very much like those made by stone tools. So, in the Americas at least, it is certainly possible that humans contributed to the extinction of the megafauna. This may also be the case in Eurasia but, so far, no direct evidence for this has been found.

A third hypothesis for the extinction of the megafauna is that they were killed by a deadly disease, perhaps one carried by humans or their dogs. But, critics ask, would a single disease be able to cause such a variety of animals to become extinct? So, despite a wealth of suggestions, the exact cause of the extinction of the Ice Age megafauna remains something of a mystery. It is entirely possible, however, that the extinctions were actually caused by a combination of different factors.

Permafrost and preservation

Although the Ice Age megafauna disappeared, it was certainly not without a trace. In fact, incredibly well-preserved remains of Ice Age animals have been found, thanks to the phenomenon of permafrost. Permafrost is permanently frozen ground, and is still found today in the arctic zones of the Northern Hemisphere, from Siberia to Alaska, Greenland and Northern Canada. Permafrost is not actually covered in ice, but in some places can be as deep as 1400 metres.

Permafrost regions have remained frozen ever since the Ice Age, and entire frozen carcasses of now-extinct Ice Age animals have been found in the permafrost of Alaska and Siberia. These animals are so well preserved that they still have skin and hair. Partially digested food has even been found in their stomachs. Because these remains are in such fine condition, these members of the Ice Age megafauna are good potential candidates for attempts to clone species that have been extinct for a long time.

Many different species have been found entombed in permafrost, including the woolly rhinoceros, horses, bison, musk ox, reindeer, wolverines, ground squirrels and ptarmigan. Some of the most spectacular finds, however, have been the remains of mammoths. Probably the best known of all the megafauna, the woolly mammoth was in fact the first species ever to be recognised as being extinct. And, if the researchers involved have their way, it will be the first extinct species to be brought back to life.

More about mammoths

The woolly mammoth bore a close resemblance to the present-day elephant but, as its name suggests, it had a very thick coat, with a shorter under layer and a very long outer layer of hair. It also had very small ears and huge tusks, with the tips pointing towards each other. The name 'mammoth' suggests they were gigantic, but in fact they were similar in size to an elephant. Siberian mammoths were even a bit smaller than elephants, with the males standing at 3 metres tall, and the females smaller again, at 2.5 metres.

Mammoths are close relatives of both African and Asian elephants. They are thought to have evolved in northern Africa about 5 million years ago. They then moved north, either across a land bridge between Africa and Europe through the Strait of Gibraltar, or perhaps via the Middle East. The mammoths that migrated to Europe spread across Europe into northern Asia, and eventually went across the Bering land bridge to North America. There were actually six species of mammoths across this range. Although mammoths evolved in Africa, by the beginning of the Ice Age there were no longer any mammoths there. The north African mammoth become extinct around 1.8 million years ago.

The ancestors of the mammoth also diversified into a number of other species. In North America, there were mastodons. In South America, there were three members of the elephant family, known as gomphotheres, which had both upper and lower tusks. And of course, there were the present-day elephant species, the African and Asian elephants, sadly the only surviving members of the whole mammoth family.

When the first mammoth remains were discovered in the 1700s, people assumed they would still be alive somewhere, hidden away in some remote corner of the globe. Although this was soon proved to be incorrect, many more mammoth remains were found and a significant amount of information has been gleaned about them. While most mammoth finds, informative though they are, are nothing more than bones and tusks, or poorly preserved carcasses, a number of finds have been truly remarkable.

In 1901, a well-preserved whole mammoth was found in Russia, way up in the arctic region. This specimen was 44 000 years old, and in a very good state of preservation when it was first discovered. Unfortunately, because of the remote location, it took three months for the excavation team with its equipment to reach the partially uncovered carcass. By the time the excavators arrived, they were dismayed to find that the carcass was no longer in such a good state of repair. The tusks had been removed and sold, and most of the head and back eaten by wolves. Not surprisingly, by this time the stench was terrible.

The excavation team found it a mammoth task (pun intended) to extract the carcass from the permafrost. In fact, it was beyond their capabilities to remove it whole, and the remains had to be extracted in pieces, which were then taken away on sleds pulled by horses. The entire process took six weeks. The mammoth was reassembled and stuffed, and displayed in the Zoological Museum in St Petersburg. Christened the Berezovka mammoth, it was the most complete mammoth ever found at the time, despite the damage that occurred before and during excavation.

Several other semi-complete or partial mammoth carcasses were discovered in Siberia over the next few decades. The most spectacular find was in 1977, when Siberian gold miner Anatoly Logachev literally hit, with his bulldozer, the remains of a baby mammoth. Only about a metre high, and probably about six months old when it died about 30 000 years ago, it was given the nickname 'Dima'. A number of scientists came to view Dima, amongst them Nikolai Vereshchagin, a mammoth expert from the Institute of Zoology in Leningrad. Unfortunately, like the Berezovka mammoth, Dima also deteriorated somewhat after extraction. On top of this, the carcass was then soaked in benzene, which caused most of its hair to fall out. Nonetheless, it was, and still is, a noteworthy specimen.

Cloning: A mammoth task

Despite the damage, Dima was so well preserved that Vereshchagin and his team decided to attempt to clone it. This was pretty extreme, especially in 1980, when no one had yet successfully cloned a living mammal. The researchers had no way of knowing whether they would succeed.

Viktor Mikhelson from Leningrad's Institute of Cytology was put in charge of the cloning project. Mikhelson and his team decided to try the following approach: first, they would look for a good adult cell from Dima—one where the cell nucleus was intact, and the DNA inside it in good condition; they would then remove the nucleus from an elephant egg, and replace it with the nucleus from Dima's cell. The embryo, if it developed, would genetically be a mammoth. It would then be implanted in a female elephant, which would, if everything went smoothly, give birth to a live mammoth calf.

Things did not go entirely according to plan. Mikhelson and his team did indeed manage to extract some DNA from Dima, which allowed them, together with Tomowo Ozawa of Nagoya University in Japan, to gain some new insights into the evolutionary relationships of mammoths. But as much as they tried, they could not find a cell which was sufficiently intact for cloning to take place. Admitting defeat, Mikhelson came to believe that it would not be possible to use mammoth cells to clone a mammoth successfully.

But this was by no means the end of the matter. Japanese scientist Kazufumi Goto did not agree with Mikhelson's assessment, and decided to continue the quest. The approach Goto decided to use, however, was slightly different to the one Mikhelson and his team had tried.

Instead of using the nucleus from an intact mammoth cell, Goto planned to use mammoth sperm to fertilise an elephant's egg. The embryo would then be implanted into a female elephant surrogate. If it survived, the resulting offspring would actually be half-mammoth and half-elephant. These hybrids could then be bred and artificially selected to produce an almost purebred mammoth, in as little as three generations. Goto could have decided to try the same approach as Mikhelson, but he thought it more likely that he would find frozen sperm than a whole intact body cell.

Goto had reason to be optimistic about the chances for success with his approach. He had already shown that dead sperm from an elephant could fertilise a live elephant egg. The embryo that resulted was implanted into a female elephant, and a live elephant calf was born. The implication was that it was not beyond the realms of possibility that mammoth sperm, dead and frozen for thousands of years, could theoretically be used to fertilise a live elephant egg.

Even before it started, however, the research was fraught with potential difficulties. It was not at all clear what sort of condition the mammoth sperm would be in, even if the carcass had been in the best possible freezing conditions. The sperm from some species is naturally more fragile than others, so would the mammoth be one of the fragile ones? Further, would all the DNA inside the sperm be intact, or would it be damaged beyond repair?

When Goto began his plans to clone a mammoth, cloning a living animal was still a very new technology. Would it be possible to extend the technology to clone an extinct animal? Although a number of cloned animals have now been produced, cloning is still not a simple process, and is more difficult in some species than others. Would the mammoth be an easy or difficult species to clone, even if the problems of using ancient tissue were overcome?

The spontaneous abortion rate is generally high for cloned embryos. Those that do survive until birth often have short life spans, due to a range of developmental defects. Would a cloned mammoth survive until birth? And how long would it live after this? How healthy would a mammoth/elephant hybrid be? An African and Asian elephant hybrid is short-lived, so would a mammoth/elephant hybrid fare any better?

Would a mammoth/elephant hybrid even be fertile? Some hybrid animals are infertile—for example the mule, a hybrid between a horse and a donkey. If a mammoth/elephant hybrid were infertile, this would make breeding a purebred mammoth from it an impossibility. Even if a mammoth/elephant hybrid were technically fertile, would this lead to a breeding population of mammoths? Elephants in captivity often stop reproducing. Would mammoths be the same?

Despite these potential problems, and the objections of other scientists who believed it would be impossible, Goto was determined to pursue the task of re-creating the mammoth.

Hunting the perfect mammoth

To achieve success, Goto knew he would need an extremely well-preserved mammoth. It would have to be a carcass which had been frozen quickly and immediately after death, before it had started to decompose. It would need to have been frozen continuously since the Ice Age, and it would still have to be frozen when the scientist found it and extracted the sperm. It would also have to be male. None of the mammoth finds so far fitted all these criteria, so it became clear to Goto that he would have to go mammoth hunting in Siberia and find his own frozen carcass.

At first, Goto had trouble finding contacts in Russia, but when he teamed up with Kazutoshi Kobayashi, a Japanese businessman, things moved more smoothly. Kobayashi had a taste for the unusual—one of his business interests at the time was importing faeces-eating flies to clean up waste from farm animals. It is not surprising that he was fascinated by the possibility of re-creating the mammoth. Importantly for Goto, Kobayashi also had contacts in Russia, which was essential for obtaining permission to go mammoth hunting. Together, the two men planned a trip to Siberia.

A number of other researchers soon joined their team. To help with the hunt itself, they recruited Pyotr Lazarev, an experienced Russian mammoth hunter who had opened a mammoth museum. Lazarev would know the best sites to look for a mammoth, and how to deal with what they found. To help with the scientific side, Akira Iritani, a reproductive biologist, joined the team.

Iritani favoured an approach similar to the original one used by Mikhelson, and thought it would be better to try to obtain a purebred mammoth straight away, by adding the DNA from a well-preserved mammoth cell to an elephant egg with the nucleus removed. Then, as Mikhelson had intended, if an embryo developed it could be implanted in an elephant. The team now had two possible approaches.

As they planned their trip to Siberia, Iritani and the rest of the team were heartened by the exciting news that Dolly the sheep had been cloned. Encouragingly, the techniques used to produce Dolly were similar to those Iritani planned to use. The signs were positive for the team. All they needed now was a good carcass.

The research team first journeyed to Siberia in 1997, but had no success. Unfazed, they made a second expedition a year later. This time, they were triumphant when they found a piece of mammoth skin, but their excitement was short-lived when they realised this was all that remained of that particular animal. Again they returned, discouraged and empty-handed.

Meanwhile, however, a different mammoth-hunting team in Siberia had happened upon another carcass. Bernard Buigues, an arctic tourism operator, was also a keen mammoth hunter, who happened to hear that a mammoth carcass had been spotted near Khatanga, on the Taimyr Peninsula. After an initial examination of the parts protruding from the permafrost, Buigues realised it was a significant discovery and called on Dirk Jan Mol, a highly knowledgeable Dutch amateur palaeontologist, to have a look.

The mammoth, which they named Jarkov, was still buried in the permafrost. The team wanted to remove the carcass whole, without defrosting it, which would be the first time this had been done. Usually, hot water is used to melt the permafrost to make extraction easier, but this destroys vital scientific information, something the team wished to avoid.

Getting the mammoth out of the ice without defrosting it was no simple matter. With great difficulty, the team finally lifted a large block of permafrost containing the mammoth by helicopter and flew it to Khatanga. They put the mammoth in an ice cave dug out of the side of a hill, where the town usually kept its food supply. It was utterly freezing, perfect for preserving the mammoth.

Would Jarkov contain sperm or body cells in good enough condition for Goto and his colleagues to attempt to clone him? It was impossible to tell at first, because the carcass was still frozen solid. The samples Goto and his team needed were not near the surface so they had to wait, agonisingly, until this part of the mammoth had been defrosted.

The researchers began to defrost Jarkov systematically, piece by piece, using hairdryers. Unfortunately, it soon became apparent that Jarkov was not going to be the answer that Goto and his team had been looking for. Jarkov, it turned out, was not nearly as intact as the team had first thought. In fact, there was hardly any body tissue left, and not even very many bones. Cloning Jarkov by any method would almost certainly be impossible. Goto and his colleagues' hopes were dashed again.

Signs of success

Finding a suitable mammoth was proving extremely difficult, which meant that cloning attempts had not yet got off the ground. But in 2003, Russian scientist Vladimir Repin made an announcement which had enormous implications for the cloning researchers. The previous summer, two frozen mammoth legs had been found in Russia, near Yakutsk. The well-preserved legs, still covered in hair, were removed from the ground, and taken in a freezer to Yakutsk's Mammoth Museum.

Repin and his team began detailed research on the legs. They saw, under a microscope, what seemed to be the answer to Goto's prayers: what appeared to be intact mammoth cells. However, further investigations have revealed that these particular mammoth legs are probably not in good enough condition for cloning. Nonetheless, the work has given mammoth cloning researchers hope that another sample will be found containing cells that are in a better state. The search for the perfect mammoth continues, and it is now a case of wait and see.

Ethics again

It may be possible one day soon to clone a mammoth. But, as with the thylacine, there are ethical issues to be considered—issues which become even more involved when talking about cloning an animal that has been extinct for many thousands of years, as the mammoth has been.

For example, no one knows what the mammoth's temperament might be like. What if it turned out to be aggressive and extremely dangerous? Most experts consider it likely that a mammoth would be similar in temperament to an elephant, because it is so closely related. But no one knows for sure, and an unknown personality is one risk of bringing back a long-extinct species. At least the thylacine's personality is not an unknown quantity, as it became extinct so recently.

Would a mammoth be able to live comfortably in a world so different to the one that existed when it was alive originally? The natural habitat of the mammoth—frozen grassy tundra—no longer exists, except for a few grasslands in Tibet. Any cloned mammoth would be forced to live in an artificial zoo environment. Is it fair to bring back an animal which will be dependent on an artificial environment? A possible solution to this issue lies in the work of a Russian ecologist, Sergei Zimov, who hopes to recreate a 'mammoth steppe' in north-east Siberia, part of a 'Pleistocene Park'. Work on this project has begun, and so far there are horses, moose, reindeer and bison in the park. These animals are removing mosses and shrubs, leaving the way clear for grasses to grow. Could this be a suitable home for a mammoth?

As with the thylacine project, there is also the question as to whether resources would be better utilised elsewhere, such as in saving endangered species. Cloning, raising and even keeping a mammoth would be an enormously expensive undertaking. With so many species currently endangered, is it ethically responsible to put such resources into a project like this? The technology involved in cloning a mammoth might be better used to clone critically endangered species to boost population numbers.

Finally, what if cloning a mammoth puts the current elephant population at risk? One theory of why the mammoths became extinct is through a deadly disease. If this were true, might the disease also be re-created and passed on to elephants? Even if this wasn't how mammoths became extinct, they still could have had diseases that elephants do not have. Some diseases can be transmitted through artificial insemination—could this lead to a disease being passed on to elephants? Cloning a mammoth is certainly not as simple as whether it is technically possible.

Conclusion: Back to the quagga

Only time will tell if cloning will succeed in bringing back any extinct species. Along with cloning research, however, a quite different approach to bringing back certain extinct species is being attempted.

Some extinct species have very close relatives still living. The ancient DNA work on the quagga, for example, confirmed it was a subspecies of the African plains zebra. Reinhold Rau, working in conjunction with game wardens, has found a number of plains zebras with a quagga-like appearance. In an exciting project which got off the ground in 1987, they are breeding these animals together with the aim of recreating the quagga.

Initially, the team captured nine zebras from the Etosha National Park, taking them to a specially designed breeding centre. A number of foals have since been born, and those with the strongest quagga-like characteristics have been selected to continue breeding.

Although they are confident that in a few years' time they will produce an animal that *looks* like a quagga, the researchers are not sure how similar it will be genetically to the original quagga. Only small portions of the quagga's DNA have been studied, so it may be impossible to tell if they have bred, in a genetic sense, a 'true' quagga. But because the original quagga was defined by its appearance, Rau and his colleagues reason that if they breed an animal that looks like a quagga, they can safely call it a quagga.

This approach, although not necessarily as dramatic as cloning an extinct animal in one step, may be the first to bring an extinct animal back to life. The project aims not just to bring back the quagga, but also to re-introduce it into the wild. This type of breeding project could not work for extinct species that are very different from other living species, such as the thylacine.

A T C G

If current research goes to plan, it may indeed be possible one day in the not so distant future to bring back a recently extinct species like the thylacine or the quagga, or maybe even an animal that has been extinct for thousands of years, like a mammoth. But just how far back could this technology potentially extend? Will it ever be possible to clone a species that has been extinct for not just thousands, but millions, of years? Could we ever do the unthinkable and make *Jurassic Park* a reality? Will we ever clone a dinosaur?

3

CRETACEOUS CAPERS
Could we really clone a dinosaur?

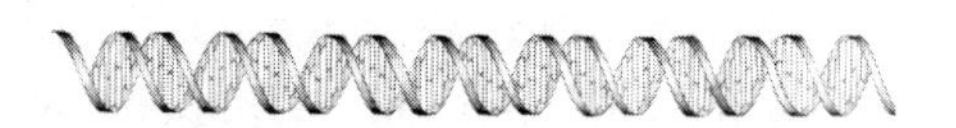

I'm not usually much of a science fiction fan, but I have to admit I loved *Jurassic Park.* In this movie, scientists working on a remote island off the coast of South America extract dinosaur blood from the guts of insects trapped in amber. They use the DNA in the blood to bring dinosaurs back to life, and the dinosaurs promptly embark on a rampage and kill a whole bunch of people and destroy lots of things. It was truly wonderful stuff. My friends and I sat spellbound throughout, jumping in fright as dinosaurs leaped out at people from behind walls and crushed them like twigs in their enormous jaws. I became so involved that when we left the theatre I half expected to see dinosaurs emerging from alleys or ambushing us from the roofs of nearby buildings.

Judging from the movie's success, and the amount of media coverage it generated, I was not the only one to be captured by it. It's hardly surprising, because dinosaurs have been a subject of public awe and fascination since their discovery in the 1800s, and the idea that they could be brought back to life in today's world was immediately seized upon with relish by scientists and the media alike.

In the first two chapters I concentrated on ancient DNA found in the remains of living things that had died anywhere from a few decades ago, such as the thylacine, to creatures that had died tens of thousands of years ago, such as mammoths and Neanderthals. DNA research has shown us that, in the right conditions, DNA can survive a remarkably long time—for many thousands of years at least.

Dinosaurs, however, became extinct not just a few thousand years ago, but about 65 million years ago. Just how long can DNA survive? Is it possible that dinosaur DNA could still exist, locked away inside bones that have been preserved for such an unimaginably long time? If dinosaur DNA does still exist, could it ever be used to create a living dinosaur?

Dinosaur discoveries

The first dinosaur fossils were discovered in the early 1800s, some 50 years before the first Neanderthal skeleton was found. Mary Anning, a young Englishwoman from Lyme Regis in coastal Dorset, often helped her father collect fossils from the beach near their home to sell to the tourists who holidayed there. Fossil collecting was then a popular hobby, although what exactly the fossils were, or how they came to be buried in layers of rock, could not be explained at the time.

The Anning family was poor, and selling fossils was a good way to supplement what would otherwise have been an inadequate income. Most of their finds were individual small bones, teeth or shells, but one day Mary's brother Joseph made a remarkable discovery. Buried in the sand was an enormous skeleton which resembled an extremely large crocodile. Mary carefully extracted the skeleton and, happily for her, managed to sell it for quite a sum.

Scientists soon heard of the find, and viewed it with great interest. Was it really some unknown type of crocodile, they wondered, or something else altogether? The significance of the find did not at first hit home. But this soon changed when, in the years following, a number of other curious, reptile-like skeletons were found. The realisation began to dawn that, at some time in the past, a diverse population of unusual reptiles had existed, a population whose members could no longer be found anywhere on Earth. In 1842, British palaeontologist Richard Owen named the group of ancient reptiles 'dinosaur', from the Greek meaning 'terrible lizard'.

Timeline: A prehistoric history

The earliest dinosaur enthusiasts did not have a clue about the true age of the skeletons they studied, nor could they know just how incredibly long the creatures that fascinated them had existed. Now known as the Mesozoic era, or 'middle life', the time of the dinosaurs began a staggering 253 million years ago with the evolution of the first dinosaur, and ended with a mass extinction around 65 million years ago.

The Mesozoic era is divided into three distinct periods: the Triassic, which began 250 million years ago and ended 210 million years ago; the Jurassic, from 210 to 140 million years ago; and the Cretaceous, which stretched from 140 to 65 million years ago.

The first dinosaurs were born on an Earth which was fundamentally different to the Earth of today. Two hundred and thirty-five million years ago, all land was clustered to form the enormous supercontinent of Pangaea, which stretched from the North Pole to the South Pole, and was surrounded entirely by one huge ocean.

Partially as a result of the different clustering of land, the climate at the time was also different. There were no ice caps in the polar areas, which had monsoonal climates. Everywhere else was hot and dry. Overall, the Earth was warmer than it is today.

It must have been a much quieter world, as there were no mammals or birds. The vegetation was also different, and was mostly made up of ferns and cycads, along with prehistoric conifers. In fact, the range of species on Earth was quite small at the beginning of the dinosaur era, as a result of a recent wave of extinctions, but this small group of species contained a group of ancient ancestral reptiles who would become the ancestors of all dinosaurs.

The dinosaur family tree

At first there were only a few small, lizard-like dinosaurs, but eventually the dinosaur family tree became more and more diverse. The earliest dinosaurs were small, light and agile, but as time went by they became steadily larger. The Jurassic period brought the enormous sauropods, which were the largest dinosaurs of all. They weighed over 70 tonnes each and could be as much as 45 metres from nose to tail. The more famous and fearsome dinosaurs, such as *Tyrannosaurus rex* and the velociraptors, were relative latecomers, appearing only near the very end of the Cretaceous period.

The dinosaurs lived on Earth for so long that they witnessed for it changing around them. First, the layout of the continents began to shift. By 155 million years ago, in the late Jurassic period, Pangaea was starting to break in two, to form Laurasia in the north and Gondwana in the south. The climate changed too. The equatorial region was still dry, but the land was becoming much wetter near the poles.

As a result of these climatic changes, the vegetation also steadily changed. Huge conifer forests—pines, cypresses and redwoods—began to diversify into all sorts of new species, some of which can still be seen today. Ferns remained common, and flowering plants made an appearance for the first time.

By the end of the Cretaceous, the last period in the dinosaur era, dinosaurs had been living on Earth for an immense 170 million years. Compare this with humans, who have existed for an estimated total of only 200 000 years, and it becomes clear just what successful species the dinosaurs really were.

The era had seen dinosaurs evolve from tiny lizard-like creatures into animals of all sizes, including immense beasts which would tower far above any species alive today. The range of shapes and sizes amongst the dinosaurs, and the ecological niches in which they lived, were extremely diverse. But the reign of the dinosaurs was to come to an abrupt end 65 million years ago when, in the blink of an evolutionary eye, they disappeared virtually without a trace.

A very sudden extinction

That the dinosaurs became extinct is a certainty. But why, after 170 million years of successful evolution, did they suddenly disappear? As with so many other events in science, this has triggered a long-running debate. These days, however, many scientists believe one huge catastrophe killed off the dinosaurs.

That catastrophe is thought to have been the impact of a huge meteorite, perhaps as big as 10 kilometres across. One day, as if out of nowhere, it slammed into the planet, creating huge clouds of dust which rose into the atmosphere and blocked the light from the Sun, plunging Earth into frigid darkness. The sudden change of climate caused plants to wither and die, in turn causing the dinosaurs, along with many other animals, to starve to death. The impact might even have triggered earthquakes or volcanic eruptions, or sent clouds of toxic gases into the atmosphere.

This theory is more than just speculation, as some compelling evidence for it has been found. Around the world there is a layer of rock, about 65 million years old, which is rich in the rare metal iridium. Iridium is not common on Earth, but is found in large quantities in meteorites—strong evidence that a large meteorite hit Earth around 65 million years ago, exactly the time the dinosaurs became extinct.

To add further weight to the theory, scientists have found large numbers of pollen grains in rocks just below the iridium-rich layer, but not many above it. This suggests that a mass extinction of plants occurred at the same time. The fact that many species of both plants and dinosaurs became extinct simultaneously also lends support to the idea that a catastrophic event killed the dinosaurs.

Scientists have even located the site where they believe the meteor struck. In the Yucatan Peninsula in Mexico, there is a rock formation which appears to be a giant crater more than 100 miles across. It is believed that this ancient crater was made by the meteor that killed the dinosaurs.

The extinction of the dinosaurs was almost total, with one important exception: the ancestors of modern birds. *Archaeopteryx* ('ancient bird') was a small dinosaur with feathered wings, dating to 150 million years ago, and is possibly the dinosaur ancestor of modern-day birds. Other non-dinosaur reptiles, including crocodiles, turtles and mammal-like reptiles, also survived. With the dinosaurs extinct, mammals began to diversify and reach a larger size, and soon came to be the dominant creatures on Earth.

Dinosaur DNA: A new frontier

Now it is time to fast-forward 65 million years, give or take a few, to the early 1990s. In a prime example of life imitating art, at around the same time *Jurassic Park* became a hit movie, real-life scientists were beginning the attempt to extract and analyse dinosaur DNA. In recent years, the relatively new field of ancient DNA research had boomed, and DNA was successfully extracted from a wide variety of extinct animals, birds and plants. Some of the samples were tens of thousands of years old, yet they still contained DNA in sufficient condition to extract and analyse.

A number of scientists were curious about just how far this technology could extend. Was there a limit to how long DNA could survive? Could it be found in the remains of living things that were not just thousands, but millions of years old? No one knew just how far back in time this exciting new technology could be pushed, but one thing was for sure, the limits were about to be seriously tested.

Supposing it could be done, why did scientists feel the need to extract DNA from such old remains? What benefits could be gained from such research? It's pretty safe to assume that a large part of the attraction of this type of work was the sheer thrill and excitement of seeing just what was possible. Pushing the frontiers of science can be challenge enough, and this was quite a frontier to be pushing. But on top of this, dinosaur DNA could actually be very useful in answering a range of scientific questions.

For example, comparing dinosaur DNA with that of present-day species could mean that we could finally know, with some certainty, which living species are most closely related to dinosaurs, a topic which has been hotly debated. Such comparisons could also confirm or refute current theories on the origins and evolutionary histories of a range of present-day species. Finally, there was the big question of whether dinosaur DNA could ever be used to clone a dinosaur.

Sources of dinosaur DNA

In *Jurassic Park* the scientists extract DNA from dinosaur blood trapped inside the guts of insects entombed in amber. In real life, scientists began to look at dinosaur fossils as potential sources of DNA. But the researchers who planned the work knew they were in for some major challenges before their real work even started.

One of the biggest barriers to the success of the research was that the process of fossilisation replaces the natural organic components of bone with inorganic minerals, including silicon and calcium. Although the fossilised bone looks much like it did before, chemically speaking it is not the same at all. The original organic material dissolves, including the DNA that was once present in it. Most dinosaur bones are fossilised through to the centre and there is no chance at all of extracting DNA from them, no matter what technology is used.

But there was some hope of success if a dinosaur bone that was not completely fossilised could be found. Such a bone might contain some of the original DNA in its non-fossilised centre. Further, the scientists thought, it might even be possible that the fossilised outer parts would have protected the inside from water and oxygen, two elements which normally cause DNA to break down rapidly. This way, they hoped, the dinosaur DNA might still be in good condition, even after millions of years.

In 1991, Mary Schweitzer, a graduate student at the Museum of the Rockies at Montana State University, made a fascinating discovery which gave considerable support to this idea. She examined a thin slice of a 65-million-year-old *Tyrannosaurus rex* bone under a microscope, and simply could not believe what she saw. There, in the centre of the ancient bone, were what appeared to be red blood cells. Was it possible, she wondered, that these cells might still have DNA in them?

Schweitzer's bone was not the only potential source of dinosaur DNA: a number of other dinosaur bones were found that were not fully fossilised and these too appeared to contain some of their original organic structures. Might they also contain dinosaur DNA?

Plant evidence

As well as these encouraging signs, those who were searching for dinosaur DNA had another source of hope. At the same time as they planned their experiments and began work, enticing reports began to surface indicating that in certain conditions it really might be possible for DNA to survive the millions of years that the dinosaur DNA researchers were hoping it could. It appeared that DNA had just been found in some plant and insect fossils that were extremely old indeed.

The first of these reports came from a team of scientists led by Edward Golenberg from the University of California. Golenberg and his team had been working with plant fossils from the Clarkia fossil beds of northern Idaho. The Clarkia beds contain large numbers of plant fossils embedded in water-soaked clay sediment which was once the bottom of an ancient lake surrounded by rainforest. The Clarkia fossils are at least 17 million years old, and have lain there, water-saturated and without oxygen, ever since the day they sank into the lake.

Golenberg and his team were most interested in 'leaf compression' fossils, which were remarkably well preserved. Despite their age, the fossilised leaves still contained intricate structures such as cell walls, and many of their organic compounds were still intact. The fossils were in such good condition that when they were extracted from the clay, they still looked green.

Given their incredible state of preservation, Golenberg was hopeful that the fossils might still contain some of their original DNA. They selected a fossilised leaf from an extinct magnolia species as their first test subject. To allow themselves the best possible chance of extracting DNA from it, they performed their initial DNA extraction experiments just minutes after removing the fossil from the sediment. Exposure to air can quickly destroy DNA, and the team wanted to ensure they had done everything possible to maximise their chances of success.

Analysis back in the laboratory led the team to the exciting conclusion that they had succeeded in their quest. The results indicated that they had found DNA which was similar, yet subtly different, to the DNA of living magnolia species. This made good biological sense, and it was reasonable to conclude the DNA they extracted was indeed from the extinct species.

For Golenberg and his team, and for the dinosaur DNA researchers, this was a very exciting result. Here was the first evidence that DNA might survive not just thousands, but millions of years. It looked as though the boundaries of ancient DNA research might indeed be pushed further than researchers had dared to hope. Although the plant fossils, at 17 million years old, were much younger than even the most recent dinosaur fossils (all of which are older than 65 million years), it was still a positive sign that dinosaur DNA might be found. And more positive signs were still to come.

Soon another American team of scientists published a second report of DNA extracted from a Clarkia fossil, this time from a leaf of an ancient species of bald cypress. A couple of years later, a French team reported that they too had managed to extract DNA from super-ancient plant compression fossils, but this time from a completely different location, at Ardèche in France. As at Clarkia, fossils from Ardèche are incredibly well preserved, and animal fossils are found there as well as a variety of plants. Ardèche fossils are slightly younger than the Clarkia fossils (about 8 million years old), but are still extremely old specimens for DNA to be found in.

There were now three reports of multi-million-year-old plant DNA, from two very different regions of the world. Before this work, the oldest ancient DNA came from 40 000-year-old mammoth remains, which is impressive enough, but nowhere near as ancient as the DNA from the plant fossils.

Insect and amber evidence

The evidence was indeed beginning to mount in support of the idea that DNA might survive far longer than anyone had initially dared to dream. Excitingly, more evidence for this was on the way, because around the same time as the plant compression fossil DNA research was being carried out, a number of other research teams were busy investigating the potential of an exciting new source of super-ancient DNA. In a scene straight out of the cinema, scientists were setting their sights on insects trapped for millions of years in amber.

In *Jurassic Park*, scientists used amber-entombed insects as a source of dinosaur DNA. The idea was that millions of years ago, an insect bit a dinosaur, drew blood, swallowed it, then became trapped in amber and died before digesting the blood. In reality, although plenty of insects preserved in amber have been found, none of them have any blood in their stomachs, dinosaur or otherwise. It was therefore impossible even to consider trying to use amber-trapped insects to obtain samples of dinosaur blood. What the scientists in real life were doing with the insects was perhaps less sensational than searching for dinosaur DNA, but every bit as interesting—they were trying to extract super-ancient DNA from the insects themselves.

Amber is the fossilised form of plant resins, secretions that harden when exposed to air. Resins have many industrial uses, for example as ingredients in paint thinners, insecticides, varnishes and polishes. The natural purpose of resin, however, is thought to be to protect the plant from insect pest attacks by trapping the insects before they can do any damage. Resin also acts as a type of biological sticking plaster for plants, forming and hardening when bark is damaged, thereby protecting the wound.

Over time, certain types of tree resin can fossilise into amber. Only a few types survive long enough for this to occur, but fossilised amber is particularly stable and can last for millions of years, particularly if it is buried underground. Amber more than 85 million years old has been found.

Most of the oldest amber found dates back to the Cretaceous period, the most recent stage of the dinosaur era. Some discoveries date back even further, to the Jurassic or Triassic periods, but amber this old is usually not as well preserved. Multi-million-year-old amber has been found in several regions of the world—North America, Lebanon, Canada, Europe and Asia—and is mined and sold in several locations worldwide.

Amber acts as a wonderful natural preservative. As the plant resin hardens, objects trapped in it are protected from decay, and are preserved incredibly well. In insects trapped in amber, for example, intricate body structures have been found to be still intact, right down to tiny, individual parts of cells—truly amazing for an insect that has been dead for up to tens of millions of years. This level of preservation is possible because the amber dehydrates and embalms the insect almost as soon as it dies.

Many things have been found trapped in amber, including grass, leaves, seeds, spiders, frogs, lizards, scorpions—even a mushroom. Most of the insects that have become trapped are species of mosquitos and sandflies. All look intact, almost lifelike.

This degree of preservation, and the recent successes in the field of ancient DNA research, made it natural that researchers would be curious as to whether the DNA in amber-entombed insects would also be preserved. A number of experts thought it highly likely, so the search began.

The extraction process

One of the researchers involved in this work was Rob DeSalle, an associate curator at the American Museum of Natural History in New York. In his 1997 book *The Science of Jurassic Park and the Lost World,* DeSalle talks in detail about his research into DNA extracted from insects in amber.

The first step in attempting to extract the DNA, DeSalle explains, is to carefully crack open the amber to allow access to the insect tissue inside. It is vital that this work be done in absolutely sterile conditions, to prevent stray DNA contaminating the experiment.

Stray DNA can be found virtually everywhere in any environment where living things have been, which of course is nearly everywhere on Earth. DNA comes from anything and everything that is, or was once, alive: living animals and the remains of dead ones, skin cells, hair, plant parts, bacteria, you name it. I can pretty much guarantee that you are sitting in an invisible sea of DNA at this very moment. Scratch your hand and release a few skin cells and you'll add to it. Pluck a hair from your head and let it fall on the floor—another drop in the DNA ocean. Sprinkle some dirt from the garden onto the carpet and you'll have added some plant DNA, bacterial DNA and a good measure of fungal DNA.

Now you've made your carpet dirty, so you'll want to vacuum. Go ahead, and in the process the vacuum cleaner will spray the stray DNA throughout every corner of your house. Go on, try it, I promise it's fun.

The vast majority of the time, this stray DNA causes no problems whatsoever for those who encounter it, which is probably why you have never before realised that this load of DNA is floating around in the environment. In ancient DNA work, however, stray DNA can be a huge problem, as it causes a phenomenon known simply as contamination.

In ancient DNA work, it is vital to ensure that only the DNA from the sample is picked up. When extraneous DNA is picked up in an experiment, the results are said to have been contaminated. This can lead to inaccurate results, and the time-consuming and often expensive experiment has to be repeated. To try to ensure this does not occur, all ancient DNA work is conducted in sterile laboratory conditions that contain as little stray DNA as possible.

So, in a sterile laboratory, our piece of amber with the insect passenger is dipped in liquid nitrogen to make it extremely brittle, and cracked open to expose the insect. Pieces of the insect can then be removed and placed in sterile, sealed test-tubes. At this stage, any DNA in the insect can be extracted and analysed, just as in any other DNA experiment.

DeSalle and a number of other researchers began to work on a range of super-ancient amber samples. DeSalle's team managed to extract small fragments of DNA from a 30-million-year-old amber-entombed termite and, later, from an amber-encased wood gnat. Other research teams also managed to extract DNA from a number of amber-entombed insects, including a bee that was between 25 and 40 million years old, and a beetle and a weevil that were both approximately 125 million years old.

Super-ancient DNA had now been found in a range of samples from two completely different types of fossils. Would the dinosaur DNA researchers be as lucky?

Now for the dinosaurs

The successful extraction of super-ancient DNA from both plant fossils and insects in amber spurred on the dinosaur DNA researchers. In fact, as is often the case in exciting new scientific research, several teams of scientists began to compete to see who could be the first to extract DNA from a dinosaur bone.

Scott Woodward, from Brigham Young University in Utah, was one of the scientists who joined the race. But before he could begin his research, he had to get his hands on a suitable dinosaur bone.

Luckily, Woodward had a geologist friend who worked in a nearby coalmine, whom he asked to keep an eye out for any bones encased in the coal. Woodward was hoping that a dinosaur bone found in coal would be a good possible source of DNA, as the coal might have protected the DNA in the bones from oxygen.

Woodward soon received the exciting news that a large skeleton had been found in the roof of the mine. The skeleton was some 80 million years old, and therefore belonged to one of the last dinosaurs in the Cretaceous period, the most recent phase of the dinosaur era. Unfortunately, the bones were so broken up that it was not possible to determine which exact species they belonged to.

Nevertheless, Woodward and his team set to work on two fragments from the skeleton. Carefully, they removed small pieces of bone from each of the fragments and examined them under a microscope. Encouragingly, the bone did not appear to be completely fossilised, and even contained structures that looked like intact cell nuclei. Importantly, there was no evidence that the organic material in the bone—including the DNA—had been replaced with inorganic minerals, which usually occurs during fossilisation.

Next, the researchers removed a number of tiny pieces from the inside of the bone fragments, ready to attempt the DNA extraction. They crushed the bone pieces to a fine powder with a glass rod and, in the same basic way as any other DNA experiment, added chemicals to extract the DNA. No doubt to Woodward's delight, the result of this work was a number of small fragments of extracted DNA.

Woodward now compared the sequences of the DNA fragments with DNA sequences from a large range of animals, birds and reptiles, all of which are stored in a large database called GenBank, a fantastic resource which is accessible via the Internet to all DNA researchers. The sequences of Woodward's DNA fragments proved to be quite different to any other mammal, bird or reptile DNA sequences. In fact, said Woodward, this DNA was 'like nothing we've ever seen before'. The unique nature of the sequences that turned up in his experiment made Woodward optimistic that he really had found dinosaur DNA, which indeed should not match DNA from any known species. A match would indicate that the DNA in Woodward's sample was not from a dinosaur at all, and that it had been contaminated with stray DNA from a present-day species.

Soon after Woodward's triumphant announcement, a team of Chinese scientists from Peking University announced that they too had successfully extracted dinosaur DNA, from the interior of a Cretaceous dinosaur egg found in Henan Province. Now there were two reports of dinosaur DNA extraction, from different teams of scientists in different parts of the world.

From Jurassic Park to reality: Dinosaur cloning

It is not every day that an announcement is made that dinosaur DNA, preserved for tens of millions of years inside an ancient bone, has been found, so it is not surprising that the news of the discovery generated a huge amount of interest. The announcement also led to the next big question on many people's minds: did this mean that *Jurassic Park* could soon resemble real life? Would it now be possible to clone a dinosaur?

Unfortunately, things are never quite as straightforward as they seem, and the fact that dinosaur DNA had been found did not automatically mean it would be possible to re-create a dinosaur. Cloning a dinosaur would by no means be as easy as it appeared in the movies.

One problem is that the developmental processes of dinosaurs are not known, and there are no species alive today that we can be certain are similar enough to dinosaurs to serve as models of a dinosaur's reproductive system. An even bigger issue, as is also the case with the thylacine cloning project, is that in order to clone a dinosaur, all the DNA from the dinosaur's genome—the complete set of genes that make up a dinosaur—would have to be known. The reports of dinosaur DNA dealt with only a tiny fraction of the DNA from the animals, meaning that cloning would be impossible unless a vastly larger amount of information could be uncovered. There was no guarantee that even a fraction of this information would still exist.

Nonetheless, the announcements created a media frenzy. Whether or not a dinosaur could be re-created, the fact that dinosaur DNA had been extracted from bones that were tens of millions of years old was truly remarkable. But amidst the elation and excitement, the voices of a number of critics were also being heard loud and clear. A storm was brewing on the horizon.

Sceptical scientists

Many scientists had been very sceptical about the possibility of extracting dinosaur DNA, questioning whether dinosaur DNA could possibly still exist, in any dinosaur bone, anywhere. One of their major objections was the fact that experimental predictions indicated that it should technically be impossible for DNA to survive anywhere near the millions of years the dinosaur DNA hunters were counting on, no matter how ideal the conditions in which the specimen was preserved.

The announcement that dinosaur DNA had been found did nothing to silence the sceptics. The problem of the expected survival time of DNA had not simply gone away. The sceptics believed there must be some other explanation for the results of the experiments.

Several independent groups of researchers—one of which included Mary Schweitzer, who had discovered red blood cell-like structures in a dinosaur bone—decided to re-analyse Woodward's results to see what they could make of them. Their results did not look good for the dinosaur DNA work, casting serious doubts on its validity. It seemed that what Woodward and his team had thought was dinosaur DNA might not be after all.

The researchers reasoned that if the DNA was from a dinosaur then, as Woodward had rightly pointed out, it should indeed be different to the DNA from any species alive today. But, although unique, any dinosaur DNA should follow the principle that it will bear a closer resemblance to DNA from more closely related species than it will to that of distant relatives. This means that dinosaur DNA should bear some resemblance to the DNA from birds and crocodiles, because these are thought to be the most closely related living things to dinosaurs. In fact, dinosaur DNA should be most similar to bird DNA, because birds are believed to be direct descendants of dinosaurs.

What the re-analysis revealed was that the 'dinosaur' DNA did not resemble most closely the DNA from birds, or even crocodiles. In fact, it was more similar to *human* DNA than to any other species. It seemed the researchers might not have been working with dinosaur DNA at all; that what they thought was dinosaur DNA was actually derived from stray human DNA, possibly even their very own.

Unsurprisingly, Woodward and his colleagues disagreed with the conclusions of the re-analysis. They argued that the relationship between dinosaurs, birds and reptiles is only a theoretical one. Just because their 'dinosaur' DNA did not fit where it would theoretically be expected to did not prove that they had not extracted genuine dinosaur DNA. They stood by their work and their original interpretation of it.

There was only one way to settle the argument, and that was for the experiments to be repeated elsewhere. For the dinosaur DNA to be accepted by everyone as genuine, it would be necessary to repeat Woodward's original experiments in an independent laboratory to see if the same results could be found, the underlying principle being that if dinosaur DNA really had been found, it would be found again by another research team.

Unfortunately, the outcome of the repeated experiments

simply strengthened the critics' position, for the DNA which had appeared in Woodward's results could not be found again.

It was not long before the other dinosaur DNA work headed in the same direction. The DNA from the Chinese dinosaur egg was also re-analysed. As with Woodward's DNA, researchers compared the Chinese 'dinosaur' DNA in more detail to a range of DNA from other living species. This time, it was clear that the DNA in the results had originated from stray DNA contamination—the 'dinosaur egg' DNA was a good match with plant and fungal DNA. It began to look increasingly likely that dinosaur DNA had not been found at all.

In a cruel twist, it was not long before questions were also raised about the other sources of super-ancient DNA. Two of the DNA experts who were sceptical of the super-ancient DNA work, Svante Pääbo and Allan Wilson, pointed out a significant potential problem with the leaf compression fossils—they had been submerged in water for millions of years.

It is a well-known fact that DNA normally degrades fairly rapidly in wet conditions; this made it difficult for Pääbo and Wilson to accept that DNA could have survived in the compression fossils, certainly not for the millions of years that these fossils had existed.

However, they were open to the possibility that the plant fossil DNA could still be genuine, saying that perhaps the pressure the fossils were under had forced all the water out of contact with the fossils and their DNA. It was a scenario they considered remote, however.

Pääbo and Wilson tried to replicate the previous team's results by extracting plant DNA from Clarkia fossils. They were unable to extract any plant DNA at all.

Finally, the DNA from insects in amber headed in the same direction. The results of the experiments could not be replicated, and the conclusion was reached that this work, too, was affected by DNA contamination.

In short, none of the experiments which found super-ancient DNA have been successfully replicated.

So what went wrong?

This was a tremendous blow for the super-ancient DNA researchers. How could such a mistake have happened? How was it possible that several teams of researchers had come to believe they had extracted DNA from ancient amber or dinosaur bones, when what they were looking at was probably nothing more than stray DNA?

In ancient DNA work, it is actually extremely easy to mistake stray DNA for genuine DNA from the sample of interest. Alan Cooper, the ancient DNA researcher I refer to in the Introduction, was one of those involved in the re-analysis of the 'dinosaur' DNA. Cooper visited New Zealand recently, and I was lucky enough to hear him speak both on the radio and at my local university about what went wrong.

As I mentioned earlier, the DNA in samples that are used for ancient DNA analysis is quite often damaged, and pieces can be missing or altered, simply because of the age of the specimen. Thus ancient DNA researchers often find themselves with very little DNA to work with. This can make research extremely difficult, because there simply isn't enough DNA to enable experiments to be conducted properly.

'The way to get around this,' explains Cooper, 'is to use a technology called PCR, or the polymerase chain reaction.' This is a technique whereby you start with a tiny amount of DNA extracted from an ancient sample and use enzymes to make literally millions of copies of it.

After the PCR process is complete, there is a huge amount of the DNA of interest to work with—in theory, all exact copies of the DNA in the original sample. It means, says Cooper, 'you've got enough there to start fooling around with'.

'But,' he adds, 'there's a slight problem with it.' The amount of ancient DNA that is present to start with is very small. But, like anywhere else, in the laboratory environment there is a great deal of stray DNA floating about: skin cells from the researchers, remnants from previous experiments, and traces from bacteria and fungi in the air. The risk is that the PCR process might accidentally copy some of this stray DNA, rather than the ancient DNA from the sample of interest.

Usually PCR is pretty accurate and zeroes in only on the piece of DNA that the researcher wants to copy, for instance the DNA in an ancient sample. But sometimes mistakes can happen and the wrong DNA is copied. When the researcher then examines the copied DNA, he or she might think it is, for example, dinosaur DNA, when in fact it may not be; rather, the sample may be contaminated by stray DNA.

What can often happen, Cooper explained, is that when a result that is obviously contaminated is obtained—for example, if the DNA is supposed to be from an extinct plant, but instead matches bacterial or human DNA—it is thrown out and the experiment is performed again. If this happens enough times, eventually researchers are virtually guaranteed, purely by chance, to get DNA that doesn't match anything they can identify. Because it is unique, it is then easy to conclude that what has been obtained is DNA from the sample of interest—however, as Cooper points out, 'by no means does that mean that it's the real thing'. As far as the super-ancient DNA work goes, Cooper believes all the results were due to contamination, not genuine super-ancient DNA.

What started out as an exciting series of ancient DNA discoveries had turned into a rather embarrassing series of mistakes. Many of those in the field of ancient DNA research were getting very annoyed, as the credibility of the entire ancient DNA field was being called into question.

As a result, calls were made for more checks to be in place before ancient DNA results were published. The first priority was to do everything possible to avoid contamination of ancient DNA experiments with stray DNA. For instance, Cooper's own laboratory, the Ancient Biomolecules Centre at Oxford University, has state-of-the-art anti-contamination measures in place: researchers change their clothes and put on special clothing before entering, and the area where DNA is extracted is isolated from the area where the PCR is carried out to avoid cross-contamination of both the sample and the PCR equipment.

But despite these measures, Cooper admits contamination is still a significant problem for ancient DNA research. Even the researchers are potential sources of contamination. 'We ourselves are incredibly dirty,' he says. We shed DNA all the time 'in sweat, on your skin, when you're breathing out, in your hair falling off, it is a major contamination issue'.

The second important requirement of all ancient DNA experiments these days is replication of results. For an ancient DNA finding to be credible, it must be repeated in an independent laboratory. If the same DNA shows up in the results both times, this is good evidence that the results are genuine.

The limited life of DNA

Researchers have now all but given up hope of ever finding DNA in remains as old as dinosaur bones. Most ancient DNA researchers now accept that, although DNA can indeed last a very long time, it does degrade eventually, especially when exposed to water and oxygen. 'If you put DNA in water and watch what happens,' says Cooper, 'you can see that it slowly breaks down.' However, he points out, the lower the temperature is, the slower the decay.

To date, the oldest genuine ancient DNA sequences have been extracted from specimens found in permafrost, in conditions which vastly slow the degradation natural in warmer climates. So just what are the temporal limits of ancient DNA thought to be? 'In permafrost conditions, we seem to find that we can go back over 100 000 years,' says Cooper. 'That's a couple of Ice Ages ago. In cold conditions, like the specimens preserved in caves and on top of mountains, it's a little bit less, somewhere around 80 000 years . . . in warm conditions, about 15 000 years . . . and in hot conditions, probably less than about 5000 years.'

And so, sadly, the current thinking is that DNA will probably never be found in the remains of dinosaur bones and eggs. Does this mean that it is absolutely impossible that a dinosaur could ever be re-created? Certainly, it suggests that a dinosaur could never be produced directly through the use of ancient DNA. But Matt Ridley, a popular science commentator, speculates that there could be a completely different method which might offer some hope.

Although dinosaurs are extinct, it is possible that many of their genes might still exist today within the DNA of birds—the apparent descendants of dinosaurs. Maybe one day, when a lot more is known about how DNA and genes work to produce characteristics in living things, it might be possible through genetic engineering to tweak the DNA of birds here and there, to emphasise the dinosaur genes remaining, and re-create a dinosaur this way. Sounds impossible? Difficult certainly, and beyond the capabilities of today's knowledge and technology, but it still leaves hope that dinosaurs may one day be re-created. But the question still remains—would, or should, we want them to be?

Ancient DNA, it seems, has now been given a time limit. Although it can be found in remains as old as the Neanderthals or Ice Age animals, it probably cannot be used to look at remains a great deal older than this, and certainly not those as old as the dinosaurs. As a result of the new criteria to ensure accuracy of ancient DNA research, the number of exceptionally ancient DNA claims has dropped away. Instead, researchers in the field have focused on more recent specimens, such as 30 000-year-old Neanderthal bones, and have been much more successful.

A T C G

As I mentioned in Chapter 2, one area in which ancient DNA researchers have achieved considerable success is the extraction of DNA from extinct species of animals, insects and birds, including my friend the moa, discussed in the next chapter. The moa is one of the most unusual flightless birds that the world has ever seen and, although a great deal has been discovered about it, for a long time there have been a number of areas of nagging uncertainty: What were its true origins? What was its relationship to the iconic New Zealand kiwi? Just how many species of moa were there? After years of frustrating uncertainty, ancient DNA research is finally providing some answers to these questions.

4

BIG BIRD

Unravelling the mysteries of the New Zealand moa

At first glance, you would not think it possible to mistake a moa for a cow. But such an improbable mistake is exactly the route by which moas first became known to science.

In 1839, a single broken moa bone was handed to the eminent English scientist Richard Owen by Dr John Rule, a London physician. Rule had been given the mysterious bone fragment by his nephew, John W. Harris, who had been living and trading in New Zealand, on the east coast of the North Island. The bone was found in a river bed and handed on to Harris, who was told by local Maori that it belonged to a large, extinct bird.

Owen was an obvious choice for Rule to give the bone to. Born in 1804, he initially trained as a surgeon but, fortuitously for science, decided instead to become an anatomist and palaeontologist. Amongst his many skills and achievements, Owen had a particular knack for spotting similarities between the anatomies of different species. It was he who first developed the idea that all vertebrates have a similar 'body plan', with different specialisations

for different lifestyles. For example, Owen recognised that the basic structure of a bird's wing, a human arm, a dog's front leg and a dolphin's flipper were the same. Owen's revolutionary work in this area would help Charles Darwin to develop the idea that these shared structures were evidence of a common ancestor—an idea of immense significance to his theory of evolution. Owen also coined the term 'dinosaur', and was a major force behind the founding of the Natural History Museum in London.

Presumably to his later embarrassment, at first Owen did not think much at all of Rule's small, broken specimen, dismissing it as a beef bone rather than a bone from a previously unknown species of giant bird. To his credit, however, Rule insisted they were looking at a specimen from a very large, and entirely unknown, extinct bird. On further examination, Owen realised it must indeed be the thigh bone of a bird—a bird the size of an ostrich, no less.

On 12 November 1839, Owen presented the bone to the Zoological Society of London, in much the same way as the first Neanderthal skeleton would be presented to science some 20 years later. His paper was the first formal declaration that New Zealand had, at some time in the past, been home to a large bird which had possibly been as big as an ostrich.

Hunting the moa

As interest in the strange bird began to grow in England, moa studies were also beginning in New Zealand. Two of New Zealand's

Opposite: *Richard Owen standing next to a Dinornis moa skeleton holding the original fragment of moa bone given to him by Dr John Rule in 1839. (Canterbury Museum, Christchurch, New Zealand)*

earliest moa experts were in fact the missionaries William Williams and William Colenso (presumably one of them was nicknamed Bill). Williams and Colenso gathered many specimens of moa bones, some of which were sent to England for further study. Moas were thus the inspiration for some of the earliest scientific endeavours in New Zealand.

Colenso also wrote a scientific paper in which he stated that the previously unnamed giant bird was called *moa* by Maoris. He also suggested that the moa could be related to the kiwi, which at the time was also rather mysterious. Because of Colenso's remote location, he had considerable trouble getting his paper published. Fortunately for him, and for moa research as a whole, the paper was eventually sent to England and passed, once again, to Richard Owen. By now convinced of the existence of the moa, Owen called for people to continue looking around New Zealand for more moas, dead or alive.

Fuelled by the interest created by these discoveries, moa hunting soon became a popular pastime in colonial New Zealand. At the time no one was completely sure whether moas were extinct, or simply very good at hiding. Different Maori tribes held different opinions on the subject: some said it was extinct, others believed it still existed in remote valleys. Moa 'sightings' were frequently reported, but none were verifiable. Adding to the debate, the first moa bone discovered was not fossilised, meaning that if the bird was extinct, it must have become so fairly recently.

This lack of knowledge about the moa is not surprising, because scientific knowledge about New Zealand's flora and fauna was at such an early stage that even the existence of the kiwi was not well documented.

Although no moas were found alive, it was not long before moa bones were found scattered throughout New Zealand. Soon, more than one moa bone was found in the one place, and complete—or almost complete—skeletons began to emerge, along with eggs. Some remains with preserved soft tissue still attached were also found.

In these early days of moa hunting, their bones were often found on or near the surface at sites that had recently been eroded, probably as the result of intense logging to create farmlands. Caves in which moa remains were discovered were also being explored, and deposits of numerous moa bones were also excavated from swamps. Finally, moa bones were found in Maori middens (rubbish dumps), indicating that they had once been hunted. Many of these remains were sent to England, and ended up with Owen.

Moa origins

It soon became obvious that moas had once existed in a wide variety of shapes and sizes and, in one form or another, had lived pretty much throughout New Zealand. It was also clear to those who studied the earliest bones that moas showed a strong similarity to a group known as the ratite family, which exists in several other regions of the world—ostriches in Africa, emus in Australia, and rheas in South America.

Early on, it was thought that the kiwi, also a member of the ratite family, must be the closest living relative of the moa. Owen, however, recognised that the moa was sufficiently different from kiwis (genus *Apteryx*) and other ratites to be placed in a new genus within the ratite family, which he named *Dinornis*, from the Greek, interpreted as 'surprising bird'.

After acknowledging that there must be a relationship between the moa and other members of the ratite family, early moa researchers pondered the question of how the group had come to be spread across such remote and distant corners of the Earth. Lacking wings, it was obvious they could not have flown. So how did they spread about? In the early days of moa hunting, this mystery simply could not be adequately explained.

It was also proving difficult to determine for certain whether moas really were extinct. Tales continued to be told of fresh moa sightings, giving some scientists hope that the birds might not be extinct after all, that one might be found alive. Others, however, were adamant that they really were extinct, and were not convinced by the tales.

Throughout the nineteenth century and into the early twentieth, the different varieties, sizes and shapes of moas were studied and divided into a range of different species by Owen and others, both in England and New Zealand. William Colenso continued his studies of the moa in New Zealand, along with his extensive botanical research. Owen himself studied New Zealand birds for over 46 years, and is responsible for naming many of the moa species.

Despite the researchers' best efforts, the early moa classifications were messy, largely because many skeletons were incomplete. A range of names was given to a huge variety of moa 'species', with classification at times based simply on one or a few bones. These species classifications were hotly debated between moa researchers, meaning there was no unified and uncontested classification scheme. The truth was that no one was sure just how many species of moa there were. Just as in the scientific arena, rumours about the moa abounded in the public domain too—about how big they were, and whether they were in fact still alive. The mysterious birds began to take on an almost mythical quality.

In the 1930s, classification of the many species of moa began in earnest. This work was given a boost in 1938 when a very large deposit of moa remains was discovered in the exotically named

Pyramid Valley Swamp, in the middle of a farm in North Canterbury in New Zealand's South Island.

As often seems to be the case, the bones were discovered by accident in rather bizarre circumstances. In a typical example of the New Zealand 'do it yourself' attitude, Joseph, the owner of the farm, and his son Rob were carrying out the grisly task of burying a dead horse in the swamp, when they happened across the moa bones. Joseph and Rob knew what they had found, but did not think the local museum would be interested. So one of the greatest scientific discoveries ever made in New Zealand was at first stored unceremoniously in a woolshed.

Eventually the bones were taken to the Canterbury Museum, Christchurch, where, of course, far from being disinterested, the scientific community was very excited by the discovery. Large-scale excavations then took place at the site, and many more moa remains were found, belonging to a range of different species. Funnily enough, no one seems to have recorded just what became of the remains of the horse.

The careful excavation at Pyramid Valley was the first time that a really large stash of moa remains had been systematically excavated by scientists. Together with a number of really good moa finds in caves in the North and South Island, the find at Pyramid Valley meant that more complete moa skeletons could be pieced together than ever before and the bird's different sizes and shapes more comprehensively classified.

Three terms that are used frequently in biological classification are species, genus and family. A *species* is a group of individuals that can interbreed with each other, but not with individuals of other species: for example, as a group, all humans can interbreed with each other, and so are members of the same species (*Homo sapiens*). Humans cannot, however, interbreed with gorillas, which are therefore a separate species. A *genus* is a group of closely related species: for example, lions (*Panthera leo*) and tigers (*Panthera tigris*) are different species, but are similar enough to be members of the same genus, *Panthera*. Finally, the next level up, *family*, is a group of closely related genera: for example, African elephants (*Loxodonta africana*), Asian elephants (*Elephas maximus*) and mammoths (*Mammuthus primigenius*), although different species and genera, are members of the same family, Elephantidae.

Over the preceding 50 or so years of moa research, it had proven surprisingly difficult to pin down just how many species, genera and families of moa there were. As time passed, moa classifications were constantly being modified and rearranged, tweaked here and there to try to reconstruct as accurately as possible just how many types of moa there were, and how they were all related to each other.

When moas were first described, it was thought that there were as many as 30 different species, a number which gradually decreased over time. This was because, in the early days, any slight variation between moa bones meant they were classified as separate species.

Following the discoveries at Pyramid Valley and other good moa sites, researchers formed the opinion that there was probably far more intra-species variation than had first been thought. For example, North Island and South Island varieties might have been a little different, but still members of the same species, a possibility not recognised by the earliest moa researchers.

As a result, the number of defined moa species was reduced further. By the late twentieth century it was thought that there might have been about eleven different species, divided into two families and six genera.

There were numerous size and shape variations between the different types of moa. The largest moa, given the species name

Dinornis giganteus, had leg bones more than a metre long, stood about 3 metres tall when upright, and could reach an estimated 270 kg in weight. My husband, a rather large man, remarked proudly that even he was much less than half the weight of this particular variety of moa. I myself can claim to be less than a quarter of the weight of a *Dinornis giganteus*—a fact of which I am also rather proud.

On the other hand, some moas were quite small. For example, *Pachyornis mappini* (Mappin's moa, named after moa enthusiast Frank Crossley Mappin, who discovered an almost complete skeleton in a cave in 1933), started from only 15 kg, about the weight of Sweep, my miniature labradoodle. In general, members of the Emeidae family were short and stocky, whereas the Dinorithidae family were tall and slender.

Moa ranges also varied considerably: *Dinornis* was found New Zealand-wide, whereas some others were only found in either the South or North Island. Their habitats also varied from mountainous regions to lowland forests to coastal shrub areas. They were a truly diverse group.

Scientists now knew a great deal about this most fascinating group of extinct New Zealand birds but, despite researchers' best efforts, a number of key issues remained, particularly regarding the relatives and origin of the moa. For example, who were the moas' closest relatives within the ratite family? It had always been assumed that New Zealand's other ratite group, the kiwis, were the closest relatives, but was this actually the case? Secondly, how did moas (and kiwis) come to be in New Zealand in the first place, so widely separated from all other ratites?

The reason these issues had been so hard to resolve was that, as no living moas were available for study, research in this area had always been forced to rely entirely on fossil evidence. Although a wealth of moa fossils had been found, they could not provide answers to these questions about the moa's eariest origins as almost all were relatively recent. Most moa bones found to date have been shown, mostly by carbon dating, to be no older than a mere 1000–3000 years, and overall it is very rare to find moa remains older than about 10 000–12 000 years.

This lack of ancient moa fossils reflects the broader situation in New Zealand palaeontology. Compared with other parts of the world, New Zealand has relatively few truly old fossils. This is due to the islands' high levels of tectonic activity (volcanic eruptions and earthquakes). This volatility in the landscape, with which New Zealanders are all too familiar, leads to uplift and subsequent erosion, processes which generally mean destruction of fossils. In addition to this inherent instability, during the Ice Ages the New Zealand landscape underwent even more upheaval due to fluctuating sea and ice levels.

As a result, it simply had not been possible to view directly the long-term changes that occurred in the fossil record throughout the evolutionary history of moas, as it is with many other species, including our own.

Evolutionary history

Based on the evidence available at the time, the most favoured theory regarding the origin of the moa was that, together with the kiwi, they had been in New Zealand as long as the country itself had existed. The theory is that when Pangaea, the enormous land mass which existed when the dinosaurs evolved, first split into two, the land that now makes up New Zealand was part of the southern super-continent of Gondwana, which also included South America, Australia, Antarctica and Africa. Eighty million years ago, not long before the end of the dinosaur era, Gondwana began to drift into pieces and the islands of New Zealand broke away. This process was actually excruciatingly slow, just a few millimetres per year, but over millions of years it amounted to a vast distance.

The flora and fauna of New Zealand, according to this theory, are believed to have drifted away from Gondwana along with the land, and so became isolated from species present on the other landmasses. The results of this long period of isolation can be seen in many of the birds, plants, insects, frogs and reptiles of New Zealand which, despite having relatives in other parts of the world, are unique.

The majority of experts believed that both moas and kiwis, being flightless, must have been amongst those species present in New Zealand at the time of the split. When Gondwana was still a single land mass, the ancestor of today's ratites was thought to have been spread over the super-continent. When the continent began to split apart, the once-united population was divided amongst the individual land masses that resulted. Over the following tens of millions of years, each isolated population began to evolve differently, leading to the different types of ratite seen today.

An alternative theory postulated that the ratites were all flighted birds at one time, that they could move from one continent to another after the continents split, and have all independently evolved flightlessness since then. This was by far the less-favoured theory, due to the complexity of the evolutionary processes that would necessarily have been involved if it were true.

The truth, however, was that due to New Zealand's scant early fossil record, no one could really know exactly which New Zealand species had been present since the split from Gondwana. Thus the origin and relations of the moa (and indeed the kiwi) remained somewhat of a mystery.

Alan Cooper, who has featured a number of times in this book, decided to see if he could find the definitive answer by using a completely different approach to the evolutionary history of the moa. Cooper is a self-confessed 'absolute rabid caver', who particularly enjoys exploratory caving, a fruitful hobby in New Zealand where, unlike in other parts of the world, there are still many caves that have never been explored. 'And of course,' says Cooper, 'they're all full of moa bones.' So many, in fact, he once said in a radio interview, that several times he used them to dig out rock falls while underground. 'They're actually quite good crowbars—quite long,' he explained sheepishly. 'Heresy now, I realise, but at the time . . .'

Cooper remembers when the first recovery of ancient DNA was reported from old bones. 'Reading this, I suddenly worked out that all these moa bones down in the caves must be full of DNA, and therefore we could start looking at New Zealand's past, at the evolution of the moa and many of the other birds and plants, the way the ecosystems changed due to the arrival of humans and the extinctions that have taken place.' Running with this idea, Cooper began to work on DNA from the moa, initially while still in New Zealand, and then at the University of California, Berkeley.

The first thing Cooper wanted to do was to use moa DNA to look at the evolution of the ratites, particularly the moa. 'They've been an evolutionary puzzle for a long time because they're flightless, and yet these giant birds are found in all the southern continents: Africa, South America, Australia, New Zealand, and there used to be one in Madagascar,' he says. 'The question is, if they're so large, and they're so flightless, how do they get to all these different land masses separated by large amounts of sea? They're certainly not going to fly—you'd need a fairly large stack of dynamite to get them off the ground.'

Cooper decided that, rather than using traditional fossil analysis to solve the mystery, he would instead attempt to extract DNA from moa bones, then compare it with other members of the ratite group. He reasoned that by looking at the similarities and differences between the DNA of the various ratite species, it might be possible to work out how long they had been separated on their different continents.

Cooper decided to try to extract DNA from a range of different genera and species of moa. The specimens he chose included *Anomalopteryx didiformis* skin and muscle held by Southland Museum in New Zealand, *Pachyornis elephantopus* muscle held by Cambridge University Zoological Museum, a *Dinornis novaezealandiae* specimen held by Yorkshire Museum, *Megalapteryx didinus* skin, rib and muscle held by the National Museum of New Zealand (now the Museum of New Zealand Te Papa Tongarewa) and *Emeus crassus* skin and tendon held by Otago Museum.

No doubt to his delight, Cooper's early experiments with DNA from these specimens were extremely successful, revealing for the first time information about the genetic characteristics of the moa. The first thing the experiments revealed was that, rather surprisingly given the large variety of shapes and sizes of the moas, the DNA from all the species tested was quite similar. All five genera tested had DNA that was almost as similar to each other as that of the three living species of kiwi are to each other. This means that despite their different appearances, moas were all quite closely related—proof indeed that appearances can be deceiving.

Next, Cooper extracted DNA from tissue samples from the living members of the ratite family, including the ostrich, cassowary, two species of rhea, the emu and, of course, the kiwi. He then compared the moa DNA with that of the living species of ratites.

These comparisons yielded a result that was quite unexpected. It had always been thought that, because of their isolation, the moa and the kiwi would be more closely related to each other than to any other members of the ratite group. To Cooper's surprise, the DNA indicated that the moa was not the kiwi's closest relative after all; in fact, the moa was no more similar genetically to a kiwi than it was to, say, an ostrich or an emu. The analysis showed that the kiwi is more closely related to the emu and cassowary of Australia than it is to the moa.

The genetic comparisons revealed that there are three distinct groups amongst the ratites: one group consisting of all the genera and species of moas; the second including all the rheas; and the third, more diverse, group consisting of the kiwi, emu, cassowary and ostrich. These results clearly indicate that the ratites must have entered New Zealand on two separate occasions—first the moas and later the kiwi. To Cooper and his colleagues, the results were quite stunning. Like everyone else, they had fully expected that the moa and kiwi were each other's closest relatives.

Cooper now believes that the moa has indeed been a part of the New Zealand fauna since Gondwana split. The kiwi, he thinks, is a more recent arrival, migrating to New Zealand after the islands had separated from Australia but were still close. It is possible that it might have 'island-hopped' across to New Caledonia and down to New Zealand during times when the land was a little more prominent from the sea. 'We do know that ratites can swim really well,' he explains.

Opposite: *The evolutionary tree of the ratite family as revealed by the study of ancient DNA. Results show that the kiwi is more closely related to the emu and cassowary than it is to the moa. The largest representative of each species has been chosen to illustrate the family tree.*

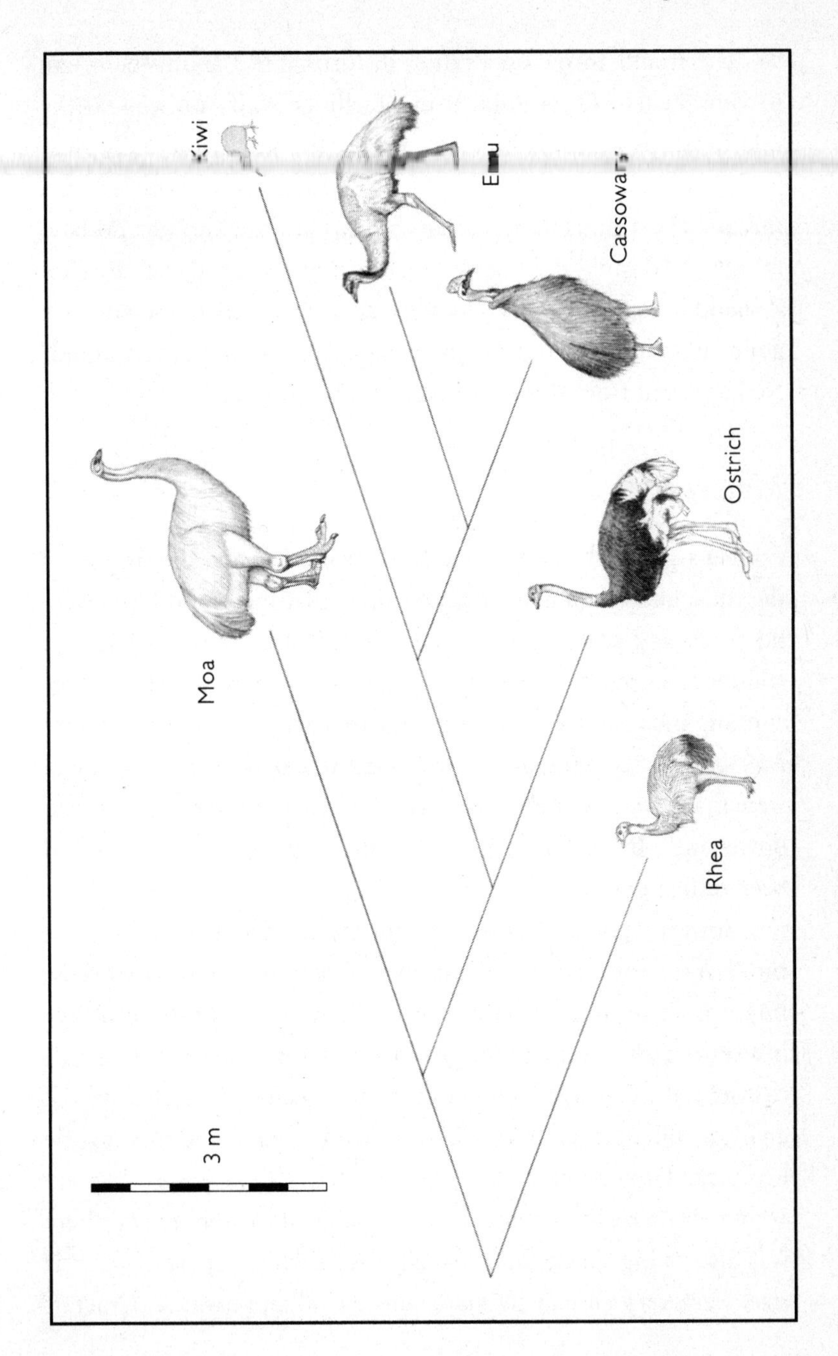
Kiwi
Ostrich
Moa
Rhea
3 m

The results throw up a rather disturbing fact about the origin of New Zealand's national icon. 'Technically, if you look at the relationships of these birds, you'd say the kiwi is actually Australian,' says Cooper. Luckily, its iconic status probably isn't under threat, as his results indicate that the kiwi has been in New Zealand for about 70 million years. With a touch of the usual friendly New Zealand–Australia rivalry, Cooper adds, 'It obviously showed very good sense and got out of Australia as soon as it possibly could.' No comment from this New Zealander about that.

Moa species

Cooper's groundbreaking moa DNA work clarified the origins of the moa, and showed, for the first time, that moas and kiwis were not as closely related as everyone had thought. But moa DNA had still more secrets to reveal. There had always been one other long-running moa mystery—just how many species of moa were there? As I mentioned previously, early moa researchers believed there were many species of moa but, over time, that number was whittled down until, by the late twentieth century, just eleven species had been settled upon.

However, researchers were by no means certain that even this modern view was correct. A number of experts believed that there might have been even fewer species than this, because of a very interesting discovery made by Joel Cracraft. In 1976, Cracraft carried out a re-classification of the moas using new techniques of analysis and, in so doing, noticed something quite odd that no one had picked up on before.

Cracraft realised that, within a particular moa genus, there was often a big species and a small species which, apart from their size, were very similar in shape and overall appearance. Cracraft suggested that perhaps the pairs of large and small moas were not different *species*, as was usually thought, but might represent different *genders* of the same species.

Size and sexual dimorphism

In nature, it is not uncommon for males and females of the same species to be different in size, a phenomenon known as sexual dimorphism. Sometimes the difference can be quite distinct, as is known to be the case in other members of the ratite family.

Based on his hunch that many species of moa might be sexually dimorphic, Cracraft described a number of moa species as different sexes, thus reducing the number of species significantly. At the time, however, he could not prove whether he was correct, for there was no way of knowing which moa remains were male and which were female. He also did not know whether the females were large and the males small, or the other way round.

David Lambert, an ancient DNA researcher and expert in bird genetics from Massey University in New Zealand, began to wonder if it would be possible to use DNA to solve the remaining moa mystery. Could he prove whether Cracraft's hunch was correct? Lambert's team was very experienced in ancient DNA work and in the analysis of bird DNA in general, which gave them confidence that their research could be successful.

The team decided to focus on a range of species representing all the major moa groups. Would it turn out that the large moas of each 'genus' were one sex, and the little ones the other? If so, which sex was big and which was small?

Technically, in order to determine the sex of the bones, all that was needed was to find a piece of DNA that was present in one gender but not the other. This is a practice commonly employed in determining the sex of human bones—for example, those obtained from archaeological sites. Human males have an X and a Y chromosome (XY), whereas human females have two X chromosomes (XX). To determine the sex of a human bone, researchers usually look for the presence of sequences of DNA that are known to be specific to the Y chromosome. There are a variety of such sequences on the human Y chromosome, and a number of these are suitable for sex determination.

In order to detect a piece of Y-specific DNA in a human bone, researchers first extract DNA from the bone, and then use the process of the polymerase chain reaction (PCR), described in the previous chapter, to make copies of the piece of Y-specific DNA. If the bone has a Y chromosome, the fragment of DNA will be copied by the PCR process, and will show up in the results of the experiment, indicating that the bone is from a male. If the bone does not have a Y chromosome, the piece of Y-specific DNA will not be copied, and will not show up in the results, indicating the bone is from a female.

Sex chromosomes in birds are the opposite to human sex chromosomes. In birds, males are the ones with two matching sex chromosomes (which, to avoid confusion, are named ZZ), and the females have different sex chromosomes (named WZ). The principle of determining the sex of bird bones in scientific research remains the same, however, and is done by looking for a piece of DNA present on one chromosome but not the other.

Lambert and his team decided that the best approach to determining the sex of moa bones would be to look for a small piece of DNA that was present only on the female W chromosome. Finding a suitable piece of moa DNA, however, was no easy task. In most species of bird there are distinct differences between male and female chromosomes, just as there are between the human X and Y chromosomes, meaning that it is easy to find a piece of DNA which can be used to distinguish males from females. In ratites, however, the male and female (W and Z) chromosomes are extremely similar, meaning that it is particularly hard to find genetic differences that distinguish males from females.

Finally, however, the team's effort led to the discovery of a suitable piece of DNA on the W chromosome that could be used to distinguish male moas from females. Once this was found, the principle of determining the sex of individual moa bones was fairly simple: DNA was extracted from each bone, and the presence of the W chromosome DNA was looked for. If present, the bird the bone belonged to must have been female; if absent, the bird must have been male.

In this way, the research team determined the sex of a large number of bones from a range of moa species. Three of these were *Pachyornis mappini* (the most important species in this study because of the large number of samples that were available), *Emeus crassus* and *Euryapteryx curtis*. Each of these three species had previously been split into two or more species, inferred from the variation present amongst their remains. However, recent moa researchers already suspected that each of these species would be sexually dimorphic and, some years before the DNA analysis was carried out, the classification was re-arranged to include just the three species. Each species, however, still contained two groups—large and small, representing males and females. Which was male or female, however, was not known before the DNA analysis.

Lambert's research revealed what others had only been able to ponder: the large moas within each genus were female, and the small ones were male.

The team also investigated the most famous genus, *Dinornis*, which contains the massive giant moas as well as a number of smaller moas. *Dinornis* became extinct about 500 years ago but before this was found throughout New Zealand.

The established view of the genus *Dinornis* was that there were three species, each a different size. These were *Dinornis struthoides*, the smallest, *Dinornis novaezealandiae*, which was medium-sized, and *Dinornis giganteus*, the whopper. By anyone's measure, the difference between the smallest and largest *Dinornis* moa was truly dramatic, with the smallest about the height and weight of a turkey, right up to the largest moas known to have existed, which were taller and more massive than ostriches.

The researchers extracted DNA from a large number of bones of each of the three types and immediately they could see that the DNA from all three was strikingly similar. So similar, in fact, that there was only one conclusion that could be reached—despite their dramatic size differences, all the *Dinornis* moas must belong to one single species.

Would it turn out that, like the other species the team had investigated, the large moas were one sex, and the little ones the other? If so, which sex was big and which was small? The research team determined the sex of a large number of *Dinornis* bones from each of the previously defined three species. Sure enough, all the smallest *Dinornis* specimens (which had been grouped into the species *Dinornis struthoides*) were males, all the huge *Dinornis* specimens (which had been grouped into the species *Dinornis giganteus*) were females, while the intermediate-sized *Dinornis* specimens (which had been grouped into the species *Dinornis novaezealandiae*) were a mixture of males and females.

The results of the research revealed what others had only been able to ponder—that the female *Dinornis* were in fact the true monsters. These findings were backed up by similar, parallel research on *Dinornis* by Alan Cooper.

Lambert is justifiably proud of his team's success in the moa DNA work. 'We understand a lot, and we feel pretty good about that, because we're in a much better position now to understand how many species of moa there were, and what their relationships were,' he says.

The future of moa research

Lambert's team is still very active in moa work and has many plans for the future. 'One of the things we're working on now is which sex actually incubated the eggs,' says Lambert. 'We think perhaps the males incubated the eggs and that females wandered around and foraged, and we're going to be able to test that.' To do this, the team will extract DNA from moa bones that were found along with eggs and, from this, determine their sex.

Another project the team is working on is to investigate moa genes that were involved in the production of colour pigments. What colour the moas were is not known for certain, because although moa feathers have been found, they were covered in clay, so their true colour could not be determined. 'We think now that we can amplify the genes that were involved in the depositing of pigment in the feathers, and we hope to be able soon to build up a picture of what colours moas were.'

The moa DNA work represents nothing short of a triumph for ancient DNA research, demonstrating how effective the technology can be in answering real scientific questions that are interesting not only in the world of research, but to society in general. After the embarrassing saga of super-ancient DNA, things were looking a little grim for the ancient DNA field for a while, but the introduction of new standards and the use of realistically aged samples have meant that ancient DNA has re-established itself as an interesting and useful field of study.

[A] [T] [C] [G]

So far we have talked about the successes (and failures) of ancient DNA research as applied to extinct species of animals, birds and human ancestors. But ancient DNA has also proved invaluable in solving more recent mysteries. One of these is the cause of a number of terrible disease epidemics, including the medieval Black Death which ravaged Europe in the fourteenth century, the epidemics of tuberculosis that broke out in the Americas after the arrival of Columbus, and the 1918 influenza pandemic. Now ancient DNA has been used to try to find the answers to some of the questions surrounding these dreadful diseases.

5

PLAGUE PROPORTIONS
Searching for the truth behind devastating historical epidemics

Ring a ring o'roses,
A pocket full of posies;
Atishoo! Atishoo!
We all fall down.
Seventeenth-century English nursery rhyme

I absolutely hate getting any type of infectious disease. Partly it's a personality thing. I'll happily admit to being a bit of a control freak, so when I get sick I don't like the feeling that somehow I'm no longer in charge of the workings of my own body. Even catching a cold makes me feel as if I have been possessed by an evil force, and I loathe the thought that some tiny microscopic life form, a virus or bacterium so small I cannot see it, has invaded my body and taken over.

Apart from my personality, I think much of the way I feel has to do with the fact that the prevention and cure of infectious diseases is now so advanced, particularly in the part of the world where I am lucky enough to live, that many of us have become used to being virtually pathogen-free most of the time, so we deeply resent it when we do become infected.

It is in fact easy to forget in today's western cultures that terrible outbreaks of disease have regularly devastated entire societies throughout history. These diseases have struck in the form of epidemics, which are outbreaks of disease that affect many people in a community at the same time, or worse, pandemics, which are epidemics that are geographically widespread, affecting a whole region or even the entire world.

Three particularly extreme examples are the medieval fourteenth-century pandemic of the Black Death, the epidemics of tuberculosis and other diseases that occurred after Columbus's voyages to the Americas and, in more recent times, the great influenza (flu) pandemic of 1918. These were three of the very worst episodes of disease in recorded history, on a scale beyond anything most of us can imagine.

Together with their deadliness and the sheer terror they caused, these outbreaks also had in common elements of utter mystery. All three have, in their own ways, baffled experts from the time they first appeared, leaving in their wake a trail of unanswered questions.

In the case of the medieval Black Death, the mystery has always been the cause of the disease. Just what infectious organism could be behind the terrible symptoms that afflicted its victims? In the case of the post-Columbus tuberculosis outbreaks in the Americas, the mystery has been in knowing who to blame. Did Columbus and his men introduce the disease into a population that had never encountered it before? Finally, in the case of the 1918 flu, the mystery lay in the severity of the disease. Why was this influenza strain so deadly when other types of flu are not?

This chapter tells the story of these three historical disease outbreaks and the secrets which, thanks to ancient DNA research, are at last being revealed.

The Black Death

One day in October 1347, a group of Italian merchant ships sailed into the port of Messina in Sicily. They had been on a voyage to the Black Sea, trading on a route linked with China. By the time the ships docked, those on board were dying from a mysterious and terrible disease. Within days the disease spread to the city and surrounding countryside: the Black Death had arrived in Europe.

In the preceding years, the Black Death had ravaged Asia. It broke out in Mongolia in the 1320s, and spread east and west from there. In China, a staggering 65 per cent of the population died.

Europeans had heard rumours of this terrible outbreak in the East but, until the arrival of the merchant ships, they had no direct experience of it. Sadly, it was not long before its effects were felt. The Black Death spread from centre to centre with frightening speed, aided by the large number of trading routes and seaports in use around Europe, as well as the river and inland routes.

The disease radiated in all directions from its point of introduction into Europe, wreaking havoc wherever it went. Less than a year after the Black Death arrived in Sicily, it reached England. By 1350, it had reached the far northern countries of Sweden and Norway, as well as Iceland and Greenland. It also spread east, into the entire Mediterranean area, and south into northern Africa. By 1352, one third of the population of Europe—some 25 million people—was dead.

The disease left death and destruction everywhere, but some cities were hit harder than others. Venice lost an estimated 90 000

A contemporary engraving depicting an English sufferer of the Black Death in 1348, showing the characteristic black boils. (Bridgeman Art Library)

people, approximately 60 per cent of its population. Milan, on the other hand, lost just 15 000 people, a mere 15 per cent. The disease appeared to be completely random: sometimes a whole family would die in the same day, and at other times one member would die one day, followed by another several weeks later. Sometimes almost everyone in an area would die; in other places only one or two. Why this occurred was a complete mystery.

SYMPTOMS

The symptoms of the Black Death were as loathsome as they were deadly. The most obvious was the black blotches caused by subcutaneous bleeding. These were accompanied by lumps or 'buboes' on the lymph nodes in the groin, armpit or neck. Buboes ranged in size from fairly small to as big as an orange, and were excruciatingly painful. As the disease progressed, the unfortunate victim would start to behave more and more irrationally until finally, up to five days later, he or she would die an agonising death.

The people of Europe simply did not know just what had hit them so hard, and finding answers proved to be extremely difficult. At the time of the pandemic, in an era before modern microscopes and laboratory techniques, the causes of disease in general were poorly understood, and the role of bacteria and viruses in particular was not yet known.

In the absence of any scientific explanation, people turned to other theories to explain the Black Death. Many sought answers in the Bible, believing the outbreak to be the vengeance of an angry God. Biblical stories told of other waves of deadly disease, including leprosy and plagues of mice. These epidemics all appeared to coincide with times of unrest and, because the Black Death had struck at an unsettled and difficult time, it seemed obvious to many that it, too, must be a manifestation of the wrath of God, a punishment for sinful behaviour and disobedience.

Far from being angry with God for his harsh punishment, the vast majority blamed themselves, and desperately tried to repent for their sins. Many behaviours were thought to have contributed to God's wrath and punishment, amongst them greed, unholy behaviour, drunkenness, and even decadent 'modern' fashions.

Another common theory was that the Black Death was the

result of a poisoned atmosphere. According to this theory, the disease was carried by a foul wind which travelled around the world, destroying those in its path. It was thought to have originated in the Indian Ocean, perhaps from water vapour bearing evil humours released by rotting fish, which would explain how the disease seemed to drift about, hitting first one area, then disappearing from it, then hitting another.

A variety of other explanations were also invoked in an attempt to understand the sudden appearance of the Black Death. Some blamed the passage of comets or the positions of the planets. Others held earthquakes and volcanic eruptions responsible for releasing evil vapours from deep inside the Earth. Yet another theory was that the disease might be the result of a medieval type of biological germ warfare, caused by an evil person brewing a deadly concoction and releasing it into the environment.

HOW DID IT SPREAD?

Because of the lack of understanding about the real cause of the disease, there was also limited knowledge of what could be done to prevent its spread. Some attempts to prevent its occurrence may even have even made the situation worse—in a desperate measure to ward off the disease, healthy individuals would swallow teeth from its victims, in the mistaken belief that this would somehow afford protection.

Some people, however, did have the insight to realise that the disease could be spread by person-to-person contact, even though the actual manner of transmission remained a mystery. It was obvious that it was not necessary to actually touch an infected or deceased person to be in danger of catching the disease, but was it passed on by an infected person's breath, or even by their gaze? Uncertainty and fear, together with the array of terrible symptoms that the unfortunate victims were afflicted with, caused other people to recoil from them in horror. Often they were left to die alone, with those around them too terrified to come near.

Along with avoiding the Black Death victims 'like the plague', other approaches were taken to try deal with the crisis. The perceived religious nature of the disease led many desperately to repent their sins. Some tried to cleanse the air of poisonous vapours. Others preferred simply to run away—and as the Black Death struck city after city, many did flee. Then there were those who decided to forgo any attempt to avoid the disease and instead lived life to the fullest, because they knew they could be struck down at any time. They indulged in food and drink, had fun at taverns and generally had as good a time as possible. Some even decided that, in the face of death and disorder, laws were suddenly irrelevant so they could do whatever they felt like, leading to considerable chaos.

HOW LONG DID IT LAST?

The pandemic raged on unabated for an agonising four years, from 1347 to 1351. The Black Death came and went in cycles which followed the seasons, spreading like wildfire in summer and early autumn, easing off in winter, then reappearing in spring. Typically, an outbreak would begin in a port, spread to the surrounding countryside and to other ports.

After the initial four-year pandemic, the situation calmed down considerably, but Europe was not yet free of the Black Death. Smaller outbreaks of the disease returned at regular intervals of six to twelve years until the mid 1600s. Finally, after a large outbreak in London in 1666, it disappeared after the Great Fire of London. After more than 300 years, the Black Death was gone, but much of its mystery remained.

WAS THE BLACK DEATH THE SAME AS THE PLAGUE?

As scientific and medical knowledge began to improve in the centuries following the Black Death pandemic, scientists tried to identify what the disease actually was. It became clear that some microorganism, bacterial or viral, was responsible, but determining just which one proved to be extremely difficult. Over time, many suggestions were made about the cause, but agreement between experts was another matter altogether. The Black Death was a surprisingly difficult disease to pin down.

By far the most common theory has been that the Black Death was a massive epidemic of the disease known simply as 'the plague'. So strong is this association that the name 'Black Death' is almost interchangeable with 'the plague' or the 'Black Plague'. But were they really the same thing?

The plague is caused by infection with the specific bacterium *Yersinia pestis.* It was first isolated in 1894, and in 1897 it was discovered that it is carried by the fleas of the black rat, *Rattus rattus.* It is now known that the bacterium can survive in rat droppings, soil and rat burrows for many years. *Yersinia pestis* is commonly spread via the bite of an infected rat flea, but when it infects the lungs of a human, as it does on occasion, the disease can spread directly from person to person through coughing and sneezing, much like a common cold.

The plague exists naturally at low levels all the time in certain areas of the world, including Uganda, the western Middle East and northern India. In fact, it still lurks in every continent except Australia. Sometimes, for reasons not entirely understood, it erupts and causes small-scale epidemics. Rarely, it spreads to become a pandemic of huge proportions, as has occurred several times in recorded history. A pandemic can last for quite some time, usually as a wave of epidemics followed by a few years of the disease disappearing, then flaring up again. Eventually, it dies back down altogether.

Why would people think that the Black Death was a massive pandemic of the plague? Partly because its symptoms appear to have been almost identical to those suffered by people who are known for certain to have suffered from the plague. In particular, people who contracted the Black Death developed buboes, a symptom that plague sufferers often develop. This common feature was first formally identified by Alexandre Yersin, a French bacteriologist, in the late 1800s.

However, despite these apparent similarities, no definite evidence could be found that the Black Death really was the plague. Complicating the matter was the fact that a number of diseases produce similar symptoms, and the plague is not the only other disease to cause buboes.

Susan Scott and Christopher Duncan, modern-day epidemiologists from Liverpool University, are particularly opposed to the idea that the Black Death was the plague, pointing out a number of flaws in this theory. First, they note that the plague is typically carried into an area by rat fleas, whereas the Black Death was able to spread in conditions where no rats, or even fleas, could survive—for example, across mountain ranges and even into rat-free Iceland. Secondly, the plague tends to spread fairly slowly, whereas the Black Death spread with breathtaking speed. Thirdly, the Black Death could be transmitted directly from person to person, whereas the plague is not usually spread this way. There *is* a rare form of the plague that can be spread from person to person, but Scott and Duncan point out that it is always fatal, and those afflicted with it are so ill, and die so quickly, that they simply cannot travel far enough to spread the disease on the scale at which the Black Death spread.

Finally, Scott and Duncan pointed out just how contagious the Black Death was, far more so than is usually the case with the plague. An outbreak of the plague in India in the 1800s killed less than 2 per cent of people in affected areas. The Black Death killed one third of the population of the areas it affected.

Scott and Duncan suggested that the Black Death could have been any one of a number of other diseases, perhaps a mutated form of a haemorrhagic virus such as Ebola or Lassa fever, or an outbreak of anthrax, typhus or even tuberculosis. Although many experts still believed that the Black Death was the plague, it could not be known for sure whose explanation was correct.

DNA AND DENTAL PULP

It was clear that new evidence was needed if the mystery of the Black Death was to be solved. In the late 1990s, Didier Raoult and his colleagues at the University of the Mediterranean in Marseilles decided to shed light on the problem by using a method quite different to anything that had been tried before.

The idea behind their proposed research was simple, yet rather ingenious. If the skeleton of a person who was known to have died from the Black Death could be found, they reasoned, they would look in those remains for DNA from the bacterium *Yersinia pestis*, which causes the plague. Finding it would be strong evidence that the Black Death and the plague were one and the same.

Fortuitously, archaeologists had recently made a discovery in Montpellier, southern France, of a medieval church cemetery containing 800 graves. When the site was excavated, four graves were identified which proved particularly interesting to Raoult and his colleagues. These graves contained many skeletons, all without shrouds, a pretty sure sign that the bodies had been buried hastily, and at a similar time.

Through a combination of carbon dating, examination of historical data, and studies of artefacts found at the site, researchers determined that the graves dated from a period during the thirteenth and fourteenth centuries. Historical records showed that the Black Death had swept through this region during this time, but no other large catastrophes had been recorded. It was likely therefore that the skeletons in the graves were indeed those of Black Death victims.

The next stage was to look for DNA from *Yersinia pestis* in the skeletons. Raoult and his colleagues decided to collect a number of teeth from one of the graves. The dental pulp inside the teeth, they reasoned, would be a good place to look for *Yersinia pestis* DNA, as dental pulp is slow to decay after death, and thus any DNA in it is likely to be well preserved. Dental pulp has the added advantage of being sealed away and thus is not as readily exposed to the problematic contamination that can occur in DNA studies, discussed in more detail in Chapter 3. Most importantly, if a person is infected with a disease such as *Yersinia pestis*, the dental pulp often becomes infected too.

In total, the team extracted 23 teeth from three different skeletons, of a man, a woman and a child. Back at the laboratory, the teeth were carefully washed before each was split in two, and the powdery dental pulp in the centre scraped out and placed in sterile test tubes.

It was then a relatively straightforward matter to test whether any DNA from *Yersinia pestis* was present in the teeth. First, as in any other DNA experiment, Raoult and his colleagues extracted all the DNA present in the dental pulp. This, of course, included human DNA as well as any *Yersinia pestis* DNA that might be present. They used the extracted DNA to set up a PCR reaction which focused specifically on trying to copy pieces of *Yersinia pestis*

DNA, while completely ignoring the human DNA. This incredible specificity is one of the real beauties of the PCR process in DNA work—not only does it produce a vast amount of copies of a DNA sequence, it can also be finely tuned to separate DNA from one species from a mixture of different DNA.

As a control, using material from the same teeth, the team set up separate PCR reactions which focused on DNA from agents that cause a number of other diseases, such as anthrax and typhus.

The results appeared to be quite conclusive. The tests for anthrax and typhus DNA were negative—meaning the victims did not die from these diseases. The PCR test for *Yersinia pestis* DNA, on the other hand, was positive. Raoult and his colleagues were understandably excited. 'We believe that we can end the controversy,' they stated when their results were announced in 2000. 'Medieval Black Death was [the] plague.'

Not everyone was as confident as Raoult and his colleagues that the mystery had been solved. Scott and Duncan were still firmly against the idea that the Black Death was caused by the plague. They said that although these tests did indeed seem to indicate that the plague was present in Europe at the right time, this did not necessarily mean that the plague *caused* the Black Death. What if there was another disease present at the same time, they asked, and that disease, not the plague, was the real killer? Raoult's results were certainly intriguing, but they were, after all, based on only one grave site. More evidence was needed before everyone would be convinced.

Thus it was that Alan Cooper, the prolific ancient DNA researcher we met earlier, decided to try to replicate the results in his own laboratory. If he could find *Yersinia pestis* DNA in the bodies of different Black Death victims, this would back up Raoult's results.

Cooper and his team set to work, analysing a large sample of 121 teeth from 66 Black Death victims collected from graves in a range of European locations, including London, Copenhagen and France.

Cooper's team carefully extracted dental pulp from the teeth of all the various victims, and tested the pulp for the presence of *Yersinia pestis* DNA, using the same basic method that Raoult had used. Far from backing up Raoult's results, however, Cooper's findings only added to the controversy. Whereas Raoult's team had found *Yersinia pestis* DNA in the teeth of all three of the victims they examined, Cooper's team found no trace of it in any of the teeth they analysed. 'We cannot rule out *Yersinia pestis* as the cause of the Black Death,' Cooper said, 'but right now there is no molecular evidence for it.'

What reason could there be for such different results? There are two possible explanations.

First, although dental pulp is better protected from contamination than many other remains, there is still the possibility that Raoult's results were somehow affected by contaminating DNA, and do not reflect the presence of genuine thirteenth- or fourteenth-century *Yersinia pestis* DNA in the teeth. Raoult, unsurprisingly, does not believe this could be the case, insisting that his team was meticulous with the measures they took to avoid this outcome.

The second possibility is that the skeletons Cooper examined did not contain any *Yersinia pestis* DNA, not because the victims did not die from the plague, but because any DNA that was present had degraded over time and could no longer be found—or, alternatively, because the victims died before the bacteria had entered their teeth.

The mystery has not yet been solved.

Columbus and tuberculosis

Although the jury is still out as to whether the Black Death was in fact the plague or not, ancient DNA has managed to shed significant light on a second historical disease mystery, this time in the Americas.

In the late fifteenth century, Christopher Columbus sailed from Europe to South America. Shortly afterwards, the people of the Americas started to die in huge numbers from dreadful outbreaks of disease. Columbus and his crew have long been held responsible for causing these huge epidemics by introducing diseases into populations that had never encountered them before.

One of the diseases that caused this devastation was tuberculosis. But was it true that the indigenous people of the Americas had never encountered it before? Was Columbus really responsible for introducing this deadly disease to a new continent?

Christopher Columbus was born in 1451 in the republic of Genoa, northern Italy, the eldest of five siblings in a family of wool merchants and weavers. The details of his early life are sketchy, but it is likely that he was brought up as a Catholic and educated entirely at home. As was common at the time, he tried his hand at a number of occupations, including wool working, but by the time he reached his mid-twenties he had discovered sailing, an occupation which would lead him to become one of the most famous explorers in history.

In about 1476, Columbus moved to Portugal, and began formulating a plan for an intrepid voyage west across the Atlantic Ocean. At the time, Europeans had a vastly different idea of the world's geography than is understood today. The known world, from a European perspective, included only Europe, Africa and Asia. These three continents were culturally and politically distinct, but there were sea and land routes connecting them, and they regularly traded with each other.

It was known that the world was spherical, but it was believed that only the area well north of the Equator was habitable. Below this was an area named the 'torrid zone', which was thought to be too hot for people to live in, although no one from Europe had actually been there. Jerusalem was considered to be the centre of the world, an idea stemming from Christianity—on maps, Europe, Asia and Africa were all centred on Jerusalem. Around the known land, a large ocean was shown.

Although it was his contact with the Americas that would make Columbus famous, it is ironic that when he planned his journey he was not trying to find the Americas at all. Columbus didn't have an inkling of the existence of another continent on the other side of the Atlantic Ocean. His actual mission was to find a route to Asia by sailing due west from Europe.

Columbus and his contemporaries had long thought that such a journey might be a possibility. They believed that the world was smaller in circumference than it actually is, which meant that there was not much ocean between Europe and Asia. They also believed that Asia projected further east than it does, and that Japan and other islands were further away from the mainland than they actually are.

Columbus experienced difficulty gaining financial support in Portugal for his proposed journey and moved to Spain in 1485 where, after a great deal of effort and persistence, he finally found the support he was looking for. King Ferdinand and Queen Isabella, although willing to support his attempt, were not at all convinced he would succeed.

Columbus arranged the use of three ships, the *Santa María*, the *Pinta* and the *Niña*, and organised crew members. The fleet set sail on 3 August 1492. They were at sea over two months, finding themselves sailing for much longer than they had anticipated. After experiencing some concern as to their whereabouts, they finally struck land.

Although they did not know it at the time, two groups of people from two vastly different parts of the world were about to meet. While Europeans were unaware of their existence, there had been people living in the Americas for at least 15 000 years.

It is thought that the first Americans arrived in the continent from Asia during the last Ice Age. Sea levels were lower during this time, and a land bridge joined Asia to America where the Bering Strait is today. Just as animals such as mammoths reached America this way, so too did people. After the last Ice Age, about 10 000 years ago, sea levels rose, the land bridge disappeared, and the Americas became isolated from the rest of the world.

Over the millennia that followed, tribes of people moved about and populated the entire American continent, from Alaska in the north to the very south of South America. Some tribes were nomadic hunter-gatherers; others stayed primarily in one place and led an agricultural lifestyle. In several areas, there were dense population centres, with empires ruling over smaller tribes.

The Vikings from Scandinavia had made contact with North America 500 years before Columbus, but they did not have much impact on the indigenous population, and other Europeans at the time of Columbus were not aware of this contact.

FIRST ENCOUNTERS

On 12 October 1492, Americans and Europeans met for the first time since the Vikings sailed across the North Atlantic, an encounter that would prove more fateful than either side could have imagined. As Columbus and his crew sailed closer to an island in what we now call the Bahamas, they could see people on the beach, watching them. They rowed ashore carrying flags and, in a manner typical of the colonial European attitude, proclaimed the land to be theirs, naming it San Salvador.

There is no record of what the local inhabitants thought of the new arrivals, but apparently they were friendly. Not realising that he had ended up in an entirely new part of the world, Columbus believed he had reached Asia. In fact, he thought he was quite near India and, because the people of the island had darker skin than Europeans, he called them Indians. They were in reality Tainos, members of the Arawak language group that inhabited an area from the Amazon to the Caribbean.

After exploring the region a little further, Columbus sailed back to Spain, still thinking he had been to Asia. Back home, he told tales of his adventures and of the people he had met and of all the things he had seen.

The new lands that Columbus had stumbled upon were a fascination and a complete mystery to people in Europe. It was quite clear that a previously unknown part of the world had been discovered, but no one was sure just how it fitted into the map of the world that they knew at the time. When people began to realise more about what exactly Columbus had found—a whole new part of the world that no one in Europe had known to exist before—he became a hero.

A year later, Columbus led a second voyage to the new lands, this time with 17 ships carrying 1200 people, the first Europeans to settle permanently in the Americas. The travellers brought horses, cows, sheep, goats, pigs and wheat, and, in a sinister sign of things to come, they also brought mounted, armed troops.

Further European voyages soon followed, with explorers venturing deeper into Central America and the South American mainland, making contact with the people there. Relationships between the indigenous Americans and the Europeans soon took a turn for the worse. Power struggles began, and the local inhabitants, who simply could not compete with the European weaponry and mounted troops, were taken captive as slaves.

The indigenous Americans were to suffer from more than the violence of war and slavery. From the time Columbus and his crew first made contact with them, they began to suffer dreadfully from a wide range of diseases, including influenza, smallpox, measles, plague, the common cold, malaria and tuberculosis.

The effects of these outbreaks were appalling. The epidemics were so severe that it is estimated between 50 and 80 per cent of the American population was killed. A whole generation of adults of child-bearing age was wiped out by disease, making it impossible for population levels to recover. Birth rates dropped and, as a result, population levels dropped even further. The organised societies that had existed before the Europeans arrived were completely devastated.

There is little doubt that these outbreaks of disease contributed to more deaths amongst the indigenous Americans than anything else. But just why did these epidemics occur? And why did they affect the Americans so badly while the Europeans remained, relatively speaking, healthy?

In populations where certain diseases have become endemic, children are usually the only ones who become infected with those diseases. Adults are almost all immune, because those who have survived the disease as children have gained immunity. The death rates amongst children in a population in which a disease is endemic can be quite high, but as the disease does not generally affect adults, the death rate amongst the population in general is quite low. However, in a population such as that in the Americas, where neither children nor adults have been exposed to diseases such as influenza or smallpox, the death rate within the population as a whole can be enormous.

This scenario undoubtedly held true for many of the diseases that the native Americans suffered from after Columbus's arrival, but the origin of one particular disease that occurred in epidemic proportions has always been open to debate: tuberculosis. Some researchers believed that tuberculosis might have been around for a long time before Columbus arrived.

This idea stemmed from the fact that a number of pre-Columbian human remains had been found with pathological effects which looked suspiciously like the result of a tubercular infection. Severe tubercular infections can cause irreversible changes in sufferers, one of the most obvious of which is the 'hunchback' spinal deformity. They can also cause lesions. However, lesions similar to those caused by tuberculosis can be caused by a variety of other diseases, including bacterial, fungal and even parasitic infections. So although the presence of lesions in these suspicious-looking human remains could indicate that the pre-Columbian people died from tuberculosis, it did not necessarily prove that this was the case.

TUBERCULOSIS AND DNA

In an attempt to solve the puzzle, pathologist Arthur Aufderheide and molecular biologist Wilmar Salo, from the University of Minnesota, together with several colleagues, decided to see if they could prove, using ancient DNA, the theory that tuberculosis had been present in the Americas before Columbus arrived.

The key to their experiments in the early 1990s was a mummified woman from southern Peru, whose body had been exhumed from a tomb several years earlier. The woman was about 40 years old when she died, and radiocarbon dating indicated that she had died no less than 1000 years ago. Crucially, her body showed signs that she could have suffered from tuberculosis. To Aufderheide and his colleagues she was the perfect candidate for testing the theory that tuberculosis had been present in the Americas centuries before Columbus was even born.

Carefully, always conscious of the threat of contamination, the researchers extracted DNA from a lung lesion and a lymph node. Using the same basic method that Raoult had used with the teeth from the suspected Black Death victims, the researchers set up a PCR reaction which zeroed in on a segment of DNA unique to the bacterium that causes tuberculosis, *Mycobacterium tuberculosis.* The results left no doubt. The woman, who died 500 years before Columbus arrived in America, *had* suffered from tuberculosis. 'This provides the most specific evidence possible for the pre-Columbian presence of human tuberculosis in the New World,' announced the research team triumphantly.

Salo, Aufderheide and their team went on to repeat the experiment with a second pre-Columbian mummy, a 12-year-old girl from Chile. They again found the presence of *Mycobacterium tuberculosis* DNA. This confirmed their first result, providing even stronger evidence that tuberculosis was not introduced into the Americas by Columbus.

Finally, wanting to get some idea of how widespread the disease had been in pre-Columbian times, American and Canadian researchers Mark Braun, Della Cook and Susan Pfeiffer looked for *Mycobacterium tuberculosis* DNA in two other sources of human remains, a fifteenth-century Canadian ossuary and an eleventh-century burial site in Mississippi. Once again, they found tuberculosis DNA, indicating that not only was tuberculosis present in pre-Columbian America, but it was also very widespread.

While ancient DNA research confirms that Columbus was not responsible for introducing tuberculosis to the Americas, the fact remains that there were devastating outbreaks of the disease amongst the indigenous populations after he arrived. Why would this be the case if tuberculosis was already endemic?

There are several possible answers. First, Columbus may have brought a different strain of tuberculosis with him, one which the American population had never encountered before, and to which they had no resistance. It is even possible that a strain of the disease arrived with Columbus and mixed with a strain already present to produce a deadly new version.

At present, one can only speculate about these scenarios. To determine whether either was the case, researchers will need to look in great detail at the DNA from the strains of tuberculosis present in Europe and in the Americas and compare it.

Another possibility is that Columbus *was* to blame for the outbreaks of tuberculosis, not because he introduced the disease, but because of the effects he and his contemporaries had on the American people. War, slavery and displacement are all factors likely to have made them more susceptible to the disease. Aufderheide suspects the epidemics were brought on in this way, by factors similar to those which cause epidemics to break out in populations of displaced people today. Behavioural and environmental changes can cause diseases that are present already in a population to suddenly break out as epidemics.

One other mystery remains. If Columbus did not introduce tuberculosis to the Americas, how did it get there? Could tuberculosis have been present in the first inhabitants? Could it have come with them when they crossed the Bering land bridge? Could it have arrived in America when the Vikings visited? Perhaps one day we will know the answers to these questions as well.

The great influenza pandemic

It is now time to move on to the story of the third huge outbreak of disease, the great influenza pandemic of 1918. By any measure, this was one of the very worst outbreaks of infectious disease witnessed in recent times. But where did it come from, and why did it spread so incredibly fast? And why, in that year, was the flu—generally a fairly benign disease to most of the population—so unusually lethal?

Researchers have been trying to find the answers to these questions for over 80 years, and now advances in ancient DNA technology are starting to uncover the truth.

The 1918 flu pandemic was like nothing ever seen before. It began in Spain in September and spread like wildfire. Within a month, the disease had reached most parts of the world, including North America as far north as Alaska, Europe, and the Pacific Islands, even the most remote areas. By October of 1918 the only places in the world where the flu had not made its mark were Australia and a number of isolated islands.

No one could doubt that what was being experienced was by far the deadliest flu epidemic in recorded history. For starters, the death rate was simply huge. Normally, the flu kills fewer than 1 in 1000 of those it infects, but the 1918 flu killed more than 1 in 40 of those unfortunate enough to become infected. Even more unusual was that most of those who died were young adults who had previously been in perfect health.

The disease affected its victims in a number of different ways. Approximately 20 per cent were lucky enough to develop no more than a regular, and not particularly severe, bout of flu, and soon recovered. The remaining 80 per cent, however, suffered forms of the disease which were much worse.

Some had an illness that for a few days seemed to be a normal type of flu, but then developed into severe pneumonia. Some recovered from this, but others were not so lucky. An unfortunate few would become extremely ill without experiencing the normal flu symptoms first. Their lungs would fill with fluid, and red spots appeared on their skin. Those who developed this form of the disease almost invariably died. A perfectly healthy person would start to feel a little ill, achy and tired, and within days or sometimes hours would be dead. The disease utterly destroyed lung tissue, meaning the victim simply suffocated.

The origin and cause of the disease was a mystery from the beginning, but rumours as to its nature were rife. Was it biological warfare? Given that the pandemic struck at the end of the First World War, it is understandable that such a notion might develop. The truth was that no one knew what it was they were facing. The disease was so different to a normal flu that it was first thought that it might be a completely different disease, perhaps botulism, cholera, typhus, or something new altogether. It eventually became clear that the illness was indeed a severe form of influenza, but its origins and the reason for its deadliness remained obscure.

A number of preventive measures were taken in an attempt to stop the spread of the disease. In the worst-affected city in North America, Philadelphia, a public campaign to discourage people from coughing, sneezing and spitting was initiated. Places where large numbers of people gathered, such as schools, pool halls and churches, were closed.

Elsewhere, other measures were adopted to prevent the disease spreading. Alcohol and antiseptic throat sprays and gargles were widely administered, and people were given masks to wear in public. In army camps, sheets were hung between beds, and down the centres of tables.

Vaccinations were trialled, but were not effective. Some contained bacteria that were thought at the time to be the cause of the flu; others contained mixtures of body fluids from flu victims. Unsurprisingly, these did nothing to prevent the disease appearing, and were rather painful as well. Realistically, there was no hope of developing a useful vaccine because no one at the time had any idea what pathogens caused even the regular strains of flu.

Just as there was no way to prevent the disease, there was also no cure for those infected. They could only be cared for and made as comfortable as possible in the hope that they would have the strength to recover.

It is estimated that 20 per cent of the world's population caught the 1918 flu. In some areas the percentage was even higher. In the US Navy, for example, 40 per cent of enlisted sailors, all young healthy adults, were infected. In Philadelphia, on 10 October 1918 alone, an incredible 759 people died. The disease was so bad that the living could not even arrange to bury the dead, because so many had died.

The pandemic raged on into 1919, killing at least 20 million people worldwide. The death toll may have even been higher, because of the absence of reporting in many countries—some estimates rate it as high as 100 million.

The 1918 flu pandemic was a great shock to western society, for improvements in medicine had meant infectious diseases overall now had much less impact than in the past. Deaths from tuberculosis, typhus and measles had been reduced greatly due to vaccinations and better hygiene and health measures in general. There had always been epidemics of diseases such as cholera and the plague throughout recorded history, and no doubt before this too. But, by 1918, it seemed that worldwide outbreaks of killer diseases should be a thing of the past.

As soon as the pandemic began, researchers were eager to discover where it came from, what caused it, and why it was so deadly. While the flu was still active, doctors attempted to discover how the disease spread by trying to infect perfectly healthy people via direct contact with those infected, or by using extracts of mucus and other body fluids. Risking the lives of healthy individuals in this way would be considered totally unethical today, but at the time it was seen as the only option to save many other people. Unfortunately, little was gained from the exercise.

Other researchers looked at the patterns of the spread of the disease to see if this would yield any clues. What they discovered only added to the mystery. This flu seemed to spread in a very strange way, appearing to jump about randomly, popping up in places that did not seem to have direct person-to-person contact with other areas with the disease.

Many other studies were made during the 1918 flu pandemic, but little was discovered that could shed any real light on what caused the disease, where it came from, or how it spread. At that stage, the agent that caused influenza was not even known, which meant there was little hope of discovering why the 1918 flu was so unusually severe.

After the pandemic passed, the disease ceased to appear even sporadically, so that it soon became impossible to directly investigate this particular flu strain, and the many questions it posed were left unanswered.

Although the 1918 flu could no longer be studied, work began in earnest to unlock the secrets of human influenza in general. The first breakthrough came in the early 1930s when the first human flu viruses were isolated, meaning that the cause of the disease could be examined. Light was also shed on the issue of how new strains might appear when it was discovered that a form of the flu virus lives in wild waterbirds, and that for reasons not yet entirely clear it can sometimes cross from birds to mammals—leading to new and sometimes severe strains in humans. It is thought that these new strains of flu might arrive in the human population by first infecting pigs, which are susceptible to bird, pig and human flus. In pigs, flus can shuffle genes, and new mixed strains pass to humans.

This research led to some suggestions about the origins of the 1918 pandemic. Perhaps the disease was the result of a bird flu strain which mutated and crossed to humans. This could have happened immediately before the pandemic struck, or even a number of years beforehand, forming a strain which might then have mutated further within the human population until finally mutating into its lethal form in 1918. Perhaps the disease passed from birds to swine, and then on to humans, or perhaps it travelled directly from birds to humans.

Without samples of the actual virus that caused the 1918 flu, it was impossible to distinguish which, if any, of these scenarios was the correct one. As to why the 1918 flu was so virulent and killed so many young people, here, too, the truth remained obscure.

THE FLU AND DNA

One day Jeffrey Taubenberger, a flu researcher at the US Armed Forces Institute of Pathology, had an interesting idea. Could genes from the 1918 flu virus still lurk in its victims' remains? If so, could this genetic information be used to reveal some of the secrets of the deadly disease? The story of the 1918 flu pandemic, and Taubenberger's work on extracting genes from the virus which caused it, is reported in detail in Gina Kolata's fascinating book *Flu.*

Taubenberger had a keen interest in solving the puzzle of the 1918 flu, viewing it as a real-life murder mystery. 'This is a detective story,' he told Kolata. 'Here was a mass murderer that was around 80 years ago who's never been brought to justice. And what we're trying to do is find the murderer.' As well as the historical interest and sheer fascination that Taubenberger so obviously felt for the topic, there was another important reason for doing the work: alarmingly, there is a real risk that such a disease might surface again. If more is known about it, there might be more hope of fighting it if it does reappear.

Taubenberger and his colleagues managed to locate three sources of lung tissue from people who succumbed to the 1918 flu. The first was from Roscoe Vaughn, a 21-year-old US Army private. The flu swept through the camp Vaughn had been sent to in South Carolina. Many men became ill and died, including Vaughn, who passed away on 26 September 1918, exactly one week after reporting sick. An autopsy had been performed, and a sample of Vaughn's infected lungs had been embedded in wax so that sections could be examined using a light microscope. No bacteria were found, but fortunately the sample was stored.

A second sample came from 31-year-old James Downs, also a US Army private, this time from a camp near New York City. Downs died just three days after being admitted to hospital. Again, an autopsy had been performed, and a sample of his bloodied, fluid-filled lungs infiltrated with wax for detailed study. Both samples had been stored from that time in the museum of the US Armed Forces Institute of Pathology.

Taubenberger and his team had a third and particularly interesting potential source of the virus. In Alaska, in a remote settlement of 80 Inuit people, the flu had swept through the village, despite its isolation, in a manner so destructive that only five adults remained alive. The dead were buried in a communal grave dug into the permafrost. The grave was opened by pathologist

Johan Hultin, who removed tissue samples and gave them to Taubenberger.

Taubenberger and his colleagues set to work on the samples to retrieve genetic material. Usually this would mean extracting the DNA. However, flu viruses, like many viruses, are a bit different in this respect from other living things because instead of using DNA as their genetic material they use a very similar molecule called RNA (ribonucleic acid). Like DNA, RNA consists of a string of smaller units, called bases. RNA is so similar to DNA, in fact, that it uses three of the same bases: adenine (A), guanine (G) and cytosine (C). The fourth base is different—instead of the base thymine (T), RNA uses a base called uracil (U). RNA actually exists in all organisms, but it is not usually the primary genetic material. Rather it is used for a variety of other functions in the cell, including the important task of transferring information from the nucleus to the rest of the cell. In viruses such as the flu virus, however, RNA replaces DNA as the genetic material, acting just like DNA does, with genes that influence how the virus is structured and how it functions. Just like DNA, flu virus RNA is also inherited when the virus reproduces.

After extracting the RNA from the flu victims' samples—in a process similar to that used to extract DNA—the researchers specifically targeted and copied a number of genes from the 1918 flu virus, using PCR. Next they worked out the sequence of bases in the copied genes. These sequences could then be compared with those of the same genes from other, more regular flu strains, and with sequences from strains of bird and swine flu. The attempt to extract and sequence flu genes from the samples was a triumph—even Taubenberger admitted he thought there was little chance of success, for the wax treatment and the workings of time had chopped the RNA into very small pieces.

The analysis of the flu gene sequences, however, only made things more confusing: when the sequences were lined up it seemed the 1918 virus had a number of similarities to bird flu, but also a number of similarities to swine and human flu. Even more strangely, the picture seemed to vary between the different 1918 virus genes themselves. Something odd was happening, but it was proving difficult to pin down just what.

After Taubenberger and his team's results were published, a team of Australian National University researchers, including Mark Gibbs and his father Adrian, together with their colleague John Armstrong, decided to re-examine the gene sequences that Taubenberger had extracted in a different way, to see if they could find a new perspective on the matter.

The ANU team decided to hone in on a particular flu gene which makes a protein known as haemagglutinin. The haemagglutinin protein protrudes from the outer surface of the flu virus particles and is required by the flu virus when it is infecting cells. Because of its role in infection, it is thought that a mutation in the gene for haemagglutinin can not only alter a virus's ability to infect, but also affect the severity of the disease it causes—its 'virulence'.

The ANU team spotted something quite unusual in the 1918 flu haemagglutinin gene, and realised they might have found the key to the mystery. They suggested that the haemagglutinin gene present in the 1918 flu virus was in fact a 'chimera'—the result of two separate haemagglutinin genes from two different strains of flu joining together. Part of the gene came from a strain of swine influenza, and part of it from a previously existing regular human influenza strain.

They developed the theory that almost immediately before the pandemic hit, a human and a swine strain of the flu somehow combined, possibly in an infected pig, and swapped pieces of their genes. The result was a strain of human flu which contained a new composite haemagglutinin.

The ANU researchers believe that this new gene was probably what made the 1918 flu strain so very infectious, leading to the enormous scale of the epidemic. The emergence of a new haemagglutinin gene may have meant that people had no resistance to the virus because their immune systems could not immediately recognise the new protein. An altered haemagglutinin gene may also have allowed the virus to infect tissues deep in the lungs, and this could have accounted for the severity of the disease.

Ironically, young people may have been particularly susceptible to this new strain of flu because their immune systems were likely to have produced a particularly strong response to it, during which, as Kolata aptly puts it, 'armies of white blood cells and fluids could rush to the lungs', leading to the pneumonia that killed so many. Their bodies may have struck back so fiercely at the unfamiliar invader that the resulting symptoms of the fight—such as fluid-filled lungs—led to their deaths. The very old and the very young may have mounted a weaker response, so that the symptoms they developed were less severe.

Thus the science of ancient DNA has helped unravel the mysteries of the 1918 flu—where it came from (biologically speaking), and why it was so infectious and so virulent. Although the conclusions are still somewhat controversial, there is little doubt that this is the best explanation so far as to the cause of one of modern history's most lethal diseases.

A T C G

One of the beauties of DNA (amongst its other endearing qualities) is that it can provide such interesting information about species or populations of animals, birds, and even infectious bacteria or viruses. But the secrets that DNA can reveal go even deeper than this. The fact that every living individual has its own unique DNA, subtly different even to other members of its own species, means that DNA can also be used to identify individuals. This is the principle behind DNA 'fingerprinting', using DNA found at crime scenes. The technology can also be used to identify bodies that are not identifiable by other means.

This feature of DNA was used to investigate two historical conundrums, the stories of which appear in the final two chapters of this book. The first is the story of Anastasia, youngest daughter of the last Romanov tsar, Nicholas II.

6

PROBLEMS OF IDENTITY
Did Anastasia survive the Russian Revolution?

A good mystery story is always intriguing, even more so when it is true. To my mind, there is no real-life mystery so famous, and so utterly intriguing, as that of the fate of Anastasia, the youngest daughter of the Russian royal family.

The effects of the Russian Revolution of 1917 still reverberate around the world. During a tumultuous series of events in the early twentieth century, Russia was transformed from an absolute monarchy, in which the tsar ruled with a power believed to have been given to him by God, into an atheist, communist republic. In the process, the Russian royal family, the Romanovs, disappeared without a trace.

The fate of the Romanovs—Tsar Nicholas, Tsarina Alexandra and their children Olga, Tatiana, Maria, Anastasia and Alexei—was shrouded in mystery from the beginning. It was generally thought they had been brutally murdered by revolutionary Bolsheviks to prevent efforts to return them to power. But the fact that their bodies could not be found meant it could not be proved just who had died or, more importantly for this mystery, whether any of them had survived.

Several years after the Russian revolution, a mysterious woman appeared in Berlin. At first she was depressed and withdrawn, and would not say who she was or where she was from. After some time, however, she confided in her nurses that she was Anastasia, the youngest of the four Romanov princesses. That she would make such a claim was not in itself considered particularly unusual, as many pretenders have taken advantage of mysterious circumstances surrounding missing aristocrats. But this time there was something different.

Anna Anderson, as she preferred to be called, bore a striking resemblance to Anastasia, and this, along with her mannerisms and demeanour, made it easy to believe she was who she claimed to be. She also had vivid memories of her childhood in the Russian royal family, and of the dreadful events that had occurred when the family disappeared. Throughout her long life, she fought to gain official recognition as Anastasia, but never succeeded, despite widespread public support.

Now, several years after her death, Anna's claims have been investigated yet again, this time with DNA evidence. The result of this remarkable work is that finally, after 60 years of mystery, her true identity can be revealed.

A brief history of the Russian Revolution

At the turn of the twentieth century, the massive eastern European country of Russia had been ruled by the Romanov family for over 300 years, its power believed to be divine, according to the Russian Orthodox faith. The current monarch was Tsar Nicholas II who, like his predecessors, was an absolute ruler. In an extremely

Portrait of the Russian royal children, taken around 1910–11. From left to right are Tatiana, Anastasia, Alexei, Maria and Olga. (Australian Picture Library)

religious country, as Russia was at the time, the right of the tsar to rule as he saw fit had historically been beyond question.

Joining Tsar Nicholas as ruler of Russia was his wife, Tsarina Alexandra Fedorovna. Alexandra was descended from both German and British royalty—her father was Grand Duke Ludwig IV of the small German principality of Hesse, and her mother was Princess Alice of Britain. As Princess Alice died when Alexandra was six, the future Russian empress was brought up by Queen Victoria, her grandmother. In an arranged marriage typical of European royal families at the time, Alexandra married the heir to the Russian throne, Nikolai (Nicholas) Aleksandrovich, in 1894.

The royal couple had five children: one son, Alexei, and four daughters, Olga, Tatiana, Maria and Anastasia.

Anastasia and her siblings were born at a turbulent time in Russian history. The divine right of the Tsar and his family to rule Russia was for the first time being questioned. Nicholas was not a strong or ruthless leader by any means, and was particularly helpless given the revolutionary mood that was brewing at the time. Slowly, his power began to erode.

Nicholas and his advisers could see trouble on the horizon, and realised they would have to take action, and soon. In an effort to defuse the situation, in 1905 the Tsar reluctantly signed the October Manifesto. This document effectively limited the Tsar's power as ruler, and allowed the creation of a constitution and a representative assembly, the Duma.

But in spite of Nicholas's allowances the unrest continued. Over the next decade political turmoil increased, as changes within the government occurred at a rapid and alarming rate. Frustration with the situation grew even stronger, and revolutionary fervour escalated. In an extreme move, a number of the Tsar's ministers and advisers were assassinated, including the now infamous family friend and adviser, Grigory Rasputin.

Crunch time came in the winter of 1917 when a bread shortage resulted in public protest and demonstrations. Although they were generally peaceful and non-violent, in a radical and desperate attempt to assert his power the Tsar ordered the army to suppress all demonstrations. Following orders, the troops opened fire on the demonstrators, with many fatalities.

The soldiers involved were deeply disturbed by this turn of events, and made a pact to disobey any further orders to shoot at civilians. Challenged by their commanding officer to obey, they turned on him and killed him. Soon there were rebellious soldiers and workers everywhere. More soldiers were sent to regain control, but they, too, deserted. The Tsar could no longer rely on his own troops to support him.

The Duma met to discuss the situation. It was obvious that something radical needed to be done before total anarchy ensued, but what that course of action should be was not so clear. The Duma debated whether to forcibly end Nicholas's leadership, but did not want to risk starting an uprising for, despite the Tsar's recent actions, the royal family still had loyal supporters amongst the people of Russia. Finally, a decision was reached—let the Tsar himself decide what to do.

The president of the Duma wrote to the Tsar and implored him to make his own concessions to save the situation. Initially, Nicholas refused to take any notice, but soon it became clear that the situation could no longer be ignored. Nicholas realised there was only one course of action: he must abdicate.

Finding a suitable replacement, however, was no easy matter. The Tsarevich, Alexei, was gravely ill with haemophilia and in no fit state to become leader of Russia. Nicholas announced to the Duma that he would abdicate in favour of his brother, the Grand Duke Mikhail Aleksandrovich. But this was not to be. Before he could take up his role as Tsar, Mikhail was warned by the Duma that if he accepted the throne his safety would be in grave danger. Fearing for his life, he turned the offer down, and Russia was left without a leader.

In the absence of a suitable candidate to become the new Tsar, the Duma was provisionally in charge of Russia. Although designed to be a democratic group, representative of all the people of Russia, in reality the Duma was dominated by professionals, wealthy landowners and industrialists. Only a few peasants were members, and it was by no means supported by the majority of the Russian people.

Perhaps inevitably, revolutionary fervour grew even stronger. A Soviet, or strike committee, was formed, whose elected members were revolutionaries. The Soviet kept a close watch on the Provisional Government, forcing it to move further to the left politically.

In March 1917, the Provisional Government placed the royal family under house arrest, supposedly for their own protection, at Alexander Palace in Tsarskoe Selo, near St Petersburg, where the family had lived since 1905. Olga, the eldest daughter, was 21 at the time, and Anastasia, the youngest, 16.

At first the Romanovs felt relatively secure, and their lives carried on more or less as before. But sensing more trouble to come, they applied for asylum in Britain. Incredibly, despite Alexandra's familial links with the British monarchy, the British Government was unwilling to antagonise the Soviet leaders and refused to grant their request.

Meanwhile, peasants seized land from landowners, and political change continued at an ever-increasing pace. Even though the Provisional Government leaned more and more to the left, there were a number of radicals who wanted to take things further, in particular the Bolshevik wing of the Marxist Russian Social Democratic Party, led by Lenin. In a bid for power, the Bolsheviks almost overthrew the Provisional Government in July.

In August, the Provisional Government decided to send the royal family into exile in Tobolsk, Siberia. This was a desperate bid to ensure its safety from the rebel forces, who wanted Nicholas and Alexandra to stand trial for crimes against the people—a trial certain to result in their execution. The move to Siberia was a closely guarded secret and even the royal family did not know where they were going. They were sent away on a train disguised as a Red Cross train, with many servants, a great deal of luggage and a guard of elite troops. The family was accommodated at the large and pleasant former governor's mansion in Tobolsk, where they stayed for eight months. They were watched over by guards, but the soldiers were pleasant and polite to them.

Meanwhile, the situation in the capital had worsened. In

November, there was a coup, this time a successful one, and the Bolsheviks, led by Lenin, overthrew the Prime Minister. This had dire consequences for the royal family. Their funds were cut off, and new soldiers were sent to watch over them who did not treat them nearly as kindly as their previous guards.

The Bolshevik influence began to spread. Even in Siberia, the royal family could not escape it. In April 1918, Nicholas, Alexandra and their daughter Maria were taken to a prison in Ekaterinburg, a mining town in the Urals. The other children were left in Tobolsk, as Alexei was not well enough to travel. On 23 May 1918, when Alexei had recovered a little, he and the three other girls were also taken to Ekaterinburg. Several servants, including Dr Eugene Sergeyevitch Botkin, the family's personal physician, were also held captive. Other loyal members of the royal entourage were either imprisoned, executed or simply cut off from all contact with the family.

The family was kept at Ipatiev House for two months, and exactly what took place during this time is still not known. The family's activities were kept a close secret from the public outside, but it is known that they were closely guarded, and their movements within their prison severely restricted.

On the night of 16 June 1918, the entire royal family—and four of their most loyal servants, including Dr Botkin—disappeared altogether, and were never seen again.

For a long time, no one really knew what had happened that terrible night. Rumours began to circulate the next day that the royal family were all dead, murdered by Bolshevik rebels, but this was initially denied by Lenin's office. Not surprisingly, there was a great deal of public interest in the fate of the royal family, both in Russia and internationally.

Soon after that fateful night, Nikolai Sokolov, a monarchist investigator, began to examine the events in detail. Based on the available evidence, which was by no means complete, he concluded that on the night of 16 June 1918, the family was herded into the cellar of Ipatiev House and ruthlessly murdered by the Bolshevik firing squad. Then, he concluded, their bodies were taken away and buried, in an unmarked and undisclosed location. Sokolov wrote a seven-volume report on the events, which has always been the official version of what happened that night.

Solokov's account, however, could not be properly verified because, despite a huge effort, the bodies of the royal family could not be found. Sokolov did find what he thought was quite likely to be the gravesite, but it contained no skeletal remains, only ashes. His conclusion was that the bodies of the royal family had been burned beyond recognition. In the absence of any recognisable remains, it was impossible to say exactly who died that night, and whether any of the family had survived.

In 1919 the Red Army took over Ekaterinburg, and Sokolov was banned from investigating the Romanovs' disappearance any further. The mystery of the fate of the Romanov family was left unsolved, but this did not mean it was forgotten. In fact, quite the contrary.

Royal rumours

After the reports of the murders, it seemed as if everyone in the western world was suddenly very interested in the Russian royal family. Books and articles about them were published, along with the contents of their diaries, as well as 'eyewitness' accounts of their lives and their murders. Newspapers and magazines were full of their photographs, along with the story of their disappearance.

But it was not just the circumstances of their deaths that gave rise to interest and speculation. Almost immediately after the royal family's disappearance, a rumour sprang up that at least one of the Tsar's daughters was still alive.

Sokolov himself added substance to the tale when he concluded that one or two of the princesses could indeed have escaped that night. Tsar Nicholas's mother, the Dowager Empress Marie Fedorovna, went even further, and insisted that her son and the rest of his family were still alive. Many others also believed that at least some members of the family were not dead, but in hiding somewhere. There were even reported sightings. A Soviet announcement that the Tsar was the only one to have died added weight to these speculations.

In the absence of any real evidence and the existence of a multitude of conflicting reports, it seemed that no one could know the truth. As far as anyone knew, it was entirely possible that a genuine member of the Russian royal family was still alive, and could appear at any time.

'Miss Unknown', the mystery woman

At nine o'clock on the night of 17 February 1920, a young woman was pulled out of the Landwhehr Canal in Berlin. Presumably attempting to take her own life, she had jumped off the Bendler Bridge into the cold water below. She was taken to hospital, where staff tried to find out who she was. Despite repeated questions, however, she refused to say who she was or why she had jumped. She had no identification that could allow authorities to trace who she was.

After several weeks in hospital, and many more attempts to get her to talk, the young woman was sent to the Dalldorf Insane Asylum, near Berlin, officially suffering from 'melancholia'. Doctors there continued to question her, but to no avail. The police searched through all the missing persons records, but no one matching her description was found. The authorities could not even establish which country she originated from, although they did not think she was German.

Christened Fräulein Unbekannt ('Miss Unknown') by the authorities, the young woman was extremely reclusive and refused to socialise with the other patients, or to take part in any of the activities at the asylum, not even daily walks. She lost a great deal of weight, and spent most of her time lying in bed, staring at the wall. She was secretive, and often hid her face under her covers.

For two years she kept silent, saying nothing about who she was. Then one day she finally spoke up and told the nurses that she was Anastasia, daughter of Tsar Nicholas II of Russia. She confided that she was frightened of being killed, and that the asylum was the only place she felt safe.

Fräulein Unbekannt's claims were not in themselves surprising, for this was not the first time since the revolution that a woman had claimed to be a member of the Russian royal family. In fact, pretenders are surprisingly common, and throughout history there have been many instances of people claiming to be long-lost members of royal or aristocratic families.

Given this long history of claimants and pretenders, it was perhaps inevitable that in the years after the Russian revolution, people claiming to be Anastasia or one of her relatives would appear. Indeed, royal family 'members' did turn up all over the world. Many were obvious frauds and were not taken seriously for any length of time. In this light, it would have been easy to dismiss Fräulein Unbekannt as simply the latest in a series of Anastasia claimants. This claimant, however, was different.

For a start, she bore an uncanny resemblance to Anastasia in particular and to the Russian royal family in general. She spoke fluent Russian, and had an intimate knowledge of the events of the Russian Revolution and of the royal family's activities. These factors made it easy for many to believe that she was who she claimed to be. Fräulein Unbekannt, instead of being dismissed as a fraud, quickly gained popular support for her claims, and soon had a collection of loyal followers.

One important group amongst Fräulein Unbekannt's supporters were exiled Russians. After the revolution, a number of Russian monarchist groups went into exile in Germany; in particular, there was a large Russian community in Berlin. After meeting her, members of these groups soon began to believe that the unknown woman could indeed be Anastasia.

Fräulein Unbekannt at first refused to cooperate with them, preferring to stay hidden. Far from putting her supporters off, her reluctance increased their belief in her. Eventually she agreed to be taken out of hospital, into the care of members of Berlin's Russian community, first staying with Baron Arthur von Kleist, who had been a provincial police officer, and his wife Maria. To them she revealed more details of her story, including her escape from the Bolsheviks and her life since then. She also said she would like to be called Anna, short for Anastasia.

Anna spent the time after her release staying with a number of different Russian exiles. During this time she suffered from a range of serious illnesses, and so was also in and out of hospital. She was still hesitant about talking to anyone about her identity and this, combined with her somewhat difficult personality, led to a number of her supporters and allies eventually becoming frustrated with her and withdrawing their support. Anna was eccentric, and had a tendency to be bad-tempered and often rather nasty to her closest allies. But as soon as one supporter gave up on her, another stepped in. The simple fact was that many people were utterly convinced that she was Anastasia.

However, there was another group of people who did not believe her story at all, in particular a number of the surviving members of the Romanov and Hesse families. The Tsarina's sister, Princess Irene of Prussia, visited Anna and was soon convinced she was not genuine. The Tsarina's brother, Grand Duke Ernst Ludwig of Hesse-Darmstadt, refused even to meet with her and was adamant she was a fraud. The Tsar's sister, Grand Duchess Olga, visited Anna, and concluded she was most certainly not Anastasia. Sydney Gibbes, who had been the Romanov children's English tutor, did not believe her either. 'She in no way resembles the true Grand Duchess Anastasia that I had known and I am quite satisfied that she is an impostor,' he said most adamantly.

Anna's biggest supporters

Despite this, a number of other relatives and friends of the Tsar and Tsarina, who had known Anastasia well, believed Anna was genuine. Some were even willing to fight for her cause. Two of these supporters were Gleb Botkin and his sister Tatiana. Gleb and Tatiana's father was Dr Eugene Botkin, the royal physician who was believed to have died alongside the Romanov family that night in 1918. As children, Gleb and Tatiana had both played with Anastasia and her brother and sisters and thus had inside knowledge of the Russian royal family.

After the revolution, Gleb had escaped to live in the United States, and Tatiana went to France. Both visited Anna, and were utterly convinced she was the Anastasia they had known as children. Gleb became her most vocal and public supporter. So convinced was he that Anna was genuine that he even went so far as to attack members of the Romanov family publicly for their denial of what he considered to be Anna's true identity.

On Gleb's suggestion, Anna moved to the United States. Gleb was a writer and published a number of articles about her, so when Anna arrived in New York there was quite a media frenzy. In an attempt to protect her identity from the media, Anna adopted the surname Anderson, which she was to keep for the rest of her life. By now it was 1928, ten years after the Russian royal family's disappearance.

After living for a number of years in the United States, Anna was sent back to an asylum in Germany, having suffered a great deal of stress and becoming inconsolably upset, largely due to the constant media attention she had had to face. Soon after her return to Germany, she gained another important supporter, Prince Frederick of Saxe-Altenburg, who was distantly related to the Romanovs.

Far from escaping attention back in Germany, Anna simply continued to become more and more famous. It seemed that everyone knew about her, whether they believed she was genuine or not. Released from the asylum, she lived for the next few years with a range of supporters and people sympathetic to her cause.

Quest for legal recognition

It was around this time that Anna began her quest to gain legal recognition of her identity as Anastasia. Although Anna genuinely wanted to be recognised as the woman she claimed to be, it was believed by many people that being recognised as the Tsar's only surviving offspring would mean more than just a family name and a historical legacy.

These people thought that if Anna really were Anastasia, she would be heir to a vast fortune. The Russian royal family was incredibly wealthy and, although most of their possessions had been confiscated during the revolution, rumour had it that the Tsar had hidden money in various places outside Russia some time before. Any surviving family members would be heir to this fortune, should it be recovered.

If it could be legally proved that Anna was Anastasia, as the only surviving daughter of the Tsar she would have been first in line, and her supporters would have been likely to gain from this too. Other Romanovs living in exile desperately needed money as well, for they too had lost everything in the revolution. In this light, it is difficult to say how many of her supporters and detractors were motivated by a true belief in who she was, or by the possible fortune involved.

The lawsuit began in 1938 and would last, on and off, for an incredible 37 years. As the court case dragged on, Anna continued to live in Germany. Aided by Prince Frederick, she went to live in a small hut in the Black Forest, in an attempt to isolate herself from the constant media attention. She stayed there for many years, surrounded by more than 60 pet cats.

The Supreme Court of West Germany finally ruled in 1970 on Anna Anderson's claim to be Anastasia. The ruling was inconclusive, the court stating that whether Anna was Anastasia could be 'neither established nor refuted'. This meant that Anna was never able to gain the recognition of her identity as Anastasia that she so desired.

However, although some of the Tsar's investments were found, the vast fortune that had been rumoured to exist never materialised. It has never been found, and it is possible that it never existed at all.

In 1968, at the age of 67, Anna returned to live in America, again at Gleb Botkin's suggestion. It was here that she met Jack Manahan, a friend of Gleb, a wealthy businessman who had an interest in the history of European royal families. He believed her story, and became her most loyal supporter. He also became her husband; she married him in December of that year. She lived the rest of her life in America with Jack, and died in 1984 at the age of 82. To the very end she swore she was the real Anastasia, a claim that could neither be proven nor refuted.

The search for the royal grave

Meanwhile, the circumstances surrounding the disappearance of the Russian royal family remained mysterious. From the beginning, the Russian Government had kept the fate of the Romanovs a closely guarded secret and, in 1928, investigations by members of the public were banned when Stalin was said to have announced, 'That's enough of this Romanov business.' No one had been allowed to investigate the royal family's disappearance since that time, but that did not stop people wondering. Rumour had been rife ever since their disappearance, and many people had come forward with various claims to have been involved, or that they somehow knew the truth.

The reality was, however, that no one, apart from those present that night, knew the real story. It was not even known exactly which members of the family had died.

In the 1970s, Russian geologist and amateur historian Alexander Avodinin decided to try to find the royal family's burial site and, with it, their remains. Risking imprisonment if caught, Avodinin began to piece together any scant information he could find that might point him in the right direction and, together with a small group of helpers, searched in secret for a burial site in the woods near Ekaterinburg.

After a number of years of fruitless effort, he was joined in his quest by Gely Ryabov, a Special Consultant to the Ministry of the Interior, who also wanted the Romanov mystery solved. Ryabov had police connections and access to Communist Party archives, which greatly helped the team in their search.

Finally, in 1979, the group discovered, in a pine forest near Ekaterinburg, a grave which they suspected could be that of the royal family. They partially excavated it, finding some remains, but re-buried them in fear of being discovered.

At first, their find was kept a closely guarded secret. But then the Soviet Union collapsed and secret archives were suddenly opened. Many previously unseen documents were found, including diaries of the royal family and eyewitness accounts of the murder and burial. The grave was excavated again, this time more thoroughly, and legally.

The Russian Government authorised an official investigation, coordinated by the Chief Forensic Medical Examiner of the Russian Federation. As a first step, traditional forensic work—facial reconstruction, age estimation, sex determination and comparison with dental records—was conducted on the remains found in the shallow grave. Nine badly damaged skeletons were pieced together. Parts of each were missing, which made the work extremely difficult.

As a result of these detailed investigations, Russian forensic authorities, together with a team of American forensic experts, came to the cautious conclusion that five of the skeletons were those of members of the Russian royal family—the Tsar, the Tsarina and three of their daughters, Olga, Maria and Tatiana. The other skeletons found with them were thought to be those of the royal physician, Dr Botkin, and three servants.

Although these forensic investigations strongly suggested that the grave was that of the Russian royal family, whether this was true could not be concluded with absolute certainty. The issue was compounded by rumours that the remains were those of another family, put there as a decoy.

DNA and the Russian royals

It was because of this uncertainty that Peter Gill, an expert from the British Forensic Science Service, was approached in 1992 with an unusual request. Would he and his colleagues be willing to begin a joint investigation with Russian experts to use DNA analysis to determine once and for all whether the remains discovered at Ekaterinburg really were those of the Russian royal family?

The premise behind the proposed DNA work was basically the same as that used in modern-day forensic DNA analysis performed when identification of a victim cannot be done by other means. In these circumstances, DNA is extracted from the victim's remains and compared with that of possible relatives, in the hope that a match can be found. In the case of the Romanovs there was a twist providing an extra challenge—not only would the analysis involve the potential identification of one of the most famous families in modern history, but it would also involve victims who had been dead for almost 80 years.

It was ironic that in order for the DNA analysis to be carried out, it was necessary to transport a number of the bones to Britain, the very country which had refused the Russian royal family asylum so many years ago. The bones were taken to Gill's laboratory, where small pieces were removed in order to extract DNA for testing. The tests involved several stages, and several angles of investigation.

First, the research team established the gender of each of the skeletons by determining which ones contained a piece of DNA that is present only on the male Y-chromosome. If this piece of

DNA was present it would indicate that the skeleton was male. Its absence would show that the skeleton was female. The results of this analysis revealed four males and five females, a result that matched perfectly with the previous forensic analysis.

Next, the researchers looked into the family relationships between the skeletons. From the previous forensic analysis, they already had an indication of who each person might be, and this was used as a starting point from which to carry out the DNA testing. To begin, the researchers compared the DNA in the skeletons they believed were those of the Tsar and Tsarina with the DNA in the three they thought were their children. The methods used in this part of the analysis were similar to those used to determine paternity in routine parental disputes involving living people. Usually, a cheek scraping is taken from the presumed father, the mother and the child. The DNA is extracted from the cheek cells, and a number of different fragments examined. The DNA profile of the child is then compared with those of the two parents. If the profiles of the father and child match, this indicates that the man is indeed the child's biological father. If two or more fragments do not fit with the father, then paternity must be excluded.

This basic method of comparing DNA from the parents with that of the children was used with the skeletons—except in this case, the DNA was extracted from the bones rather than from cheek scrapings. The results showed that there was indeed a family relationship between the five suspected Romanov skeletons.

The final stage in the DNA work was to investigate whether the remains really were those of the Russian royal family. The work so far had shown that there was definitely a family group involved, but it did not prove they were the missing Romanovs. As is often the case in forensic DNA analysis, it was necessary to locate a living relative with whose DNA the DNA in the skeletons could be compared. In this case, the living relative most suited to the job was none other than Britain's Prince Philip, the husband of Queen Elizabeth II.

Prince Philip and Tsarina Alexandra are blood relatives, both direct descendants of Queen Victoria. Prince Philip is in fact a grand-nephew of the Tsarina and thus is a blood relative of her children as well. This made him a perfect candidate for comparison. Obviously as keen to know the truth as any, Prince Philip generously provided a blood sample for the purpose.

Some results at last

Gill and his colleagues extracted DNA from Prince Philip's blood sample and compared it with DNA extracted from the bones thought to be those of the Tsarina and her three children. As they had hoped, it was a perfect match. This remarkable piece of analysis finally made it possible to confirm that four of the bodies found in the woods near Ekaterinburg were indeed those of the Russian Tsarina Alexandra and three of her daughters.

There was now one further piece of analysis to perform, to verify that the skeleton thought to be that of Tsar Nicholas himself was genuine. Finding a living relative of Nicholas willing to provide a blood sample proved much more difficult than it had been to find a relative of the Tsarina, but eventually the researchers were able to locate two of his relatives willing to donate samples of their DNA: a great-great-grandson and a great-great-great-

Opposite: *Family tree of the Russian royal family, showing the blood relationship to Prince Philip, Duke of Edinburgh and consort of Queen Elizabeth II. Prince Philip's grandmother, Princess Victoria of Hesse, was the sister of Tsarina Alexandra, and his mother was Anastasia's cousin.*

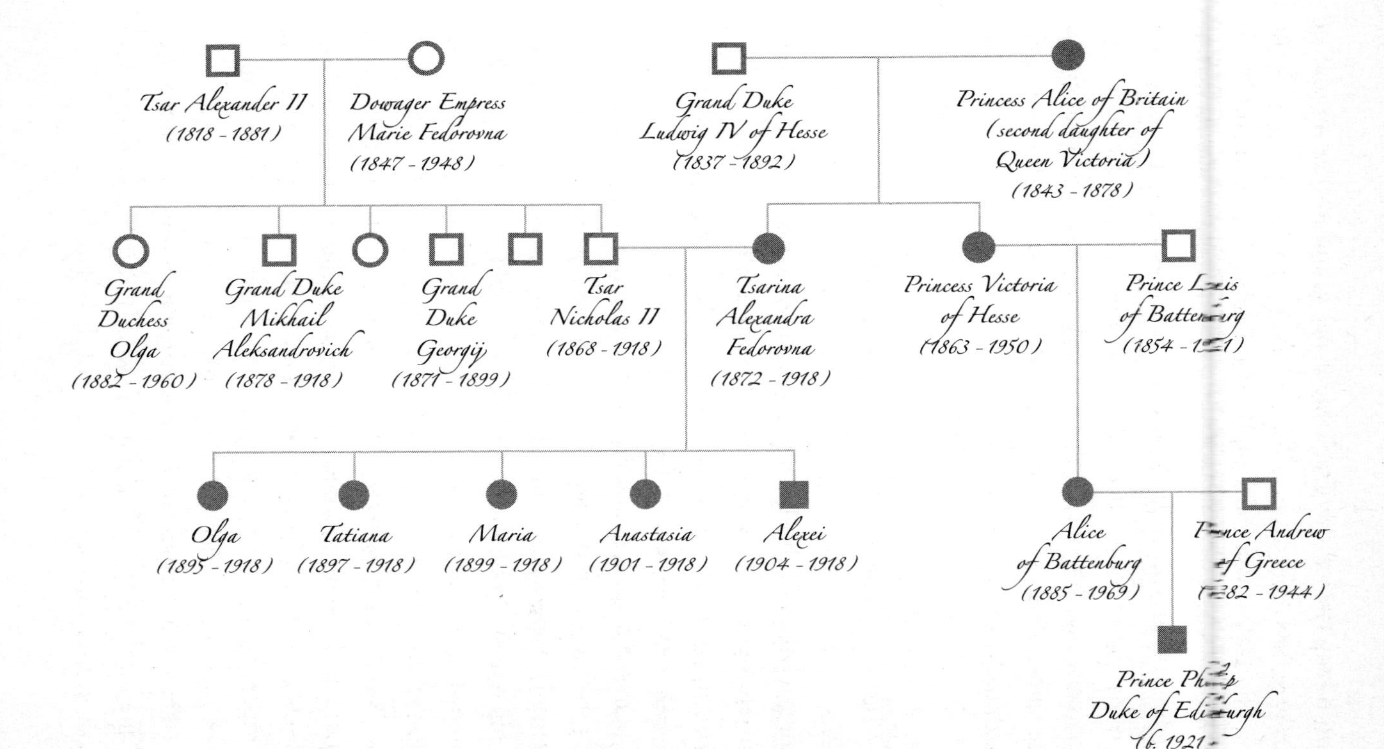
Tsar Alexander II (1818 - 1881)
Dowager Empress Marie Fedorovna (1847 - 1948)
Grand Duke Ludwig IV of Hesse (1837 - 1892)
Princess Alice of Britain (second daughter of Queen Victoria) (1843 - 1878)
Grand Duchess Olga (1882 - 1960)
Grand Duke Mikhail Aleksandrovich (1878 - 1918)
Grand Duke Georgij (1871 - 1899)
Tsar Nicholas II (1868 - 1918)
Tsarina Alexandra Fedorovna (1872 - 1918)
Princess Victoria of Hesse (1863 - 1950)
Olga (1895 - 1918)
Tatiana (1897 - 1918)
Maria (1899 - 1918)
Anastasia (1901 - 1918)
Alexei (1904 - 1918)
Alice of Battenburg (1885 - 1969)

granddaughter of the Tsar's maternal grandmother, Louise of Hesse-Cassel.

The DNA from the samples the Tsar's relatives provided was compared with the DNA in the skeleton thought to belong to Tsar Nicholas. Unfortunately, the result was not as straightforward as it was for Alexandra and her children. The DNA in the skeleton was a very close match to that of the Tsar's relatives, but it was not *exactly* the same. This meant that although the researchers were fairly confident that the skeleton was that of the Tsar, they could not say so with absolute certainty. DNA from a closer relative would be necessary to confirm the result.

But where could such a relative be found? The researchers had already experienced more than a little difficulty in locating any living relatives of the Tsar at all, let alone an extremely close one. But an ingenious solution was at hand. Instead of looking for a close living relative of the Tsar, they decided they would look instead towards one who had already died.

One of Tsar Nicholas's brothers, Grand Duke Georgij, had died of tuberculosis at the age of 28. Fortuitously for the researchers, Georgij's body was entombed in the family vault in St Petersburg. If his remains could be exhumed, this would provide a perfect, if somewhat unpleasant, opportunity to obtain DNA from a very close relative.

The researchers were able to obtain permission to exhume Grand Duke Georgij's body in 1994 and remove a sample from which to extract DNA. It was then a fairly simple matter to compare Georgij's DNA with the DNA from the skeleton thought to be that of Tsar Nicholas. This time, it was a perfect match.

The research, said Gill, 'proves virtually beyond doubt that five of the nine skeletons found in the woods near Ekaterinburg were those of the Tsar, Tsarina and three of their daughters'.

The missing children

The mystery of what really became of the Russian royal family could almost be laid to rest. Almost, that is, but for one thing. As well as identifying who *was* in the grave, the forensic and DNA analysis had shown who *wasn't* there—the two youngest children, Alexei and Anastasia.

There were two possible explanations: either Anastasia and Alexei had died and were buried elsewhere, or they had somehow managed to escape. If the second scenario was correct, could this mean that Anna Anderson had been telling the truth all along? Would this be the evidence she had waited all her life for?

There was only one way to know if Anna had been genuine, and that was to compare her DNA with that of the Romanov family. Dick Schweitzer decided to finance the necessary research. Schweitzer's wife Marina is the granddaughter of Dr Eugene Botkin, personal physician to the Romanovs, and the daughter of Gleb Botkin, one of Anna Anderson's most fervent supporters. Marina and Dick were eager to have the DNA analysis carried out. 'All I want is the truth,' said Schweitzer.

Schweitzer commissioned a team of DNA researchers, led once again by Peter Gill. Before the research could go ahead, it was crucial that a sample of Anna's DNA be located for comparison with the DNA from the Romanov family. If Anna really was Anastasia, her DNA would be a perfect match with that of Anastasia's mother, Alexandra, and her sisters.

Locating some of Anna's DNA proved to be the first major challenge, for when Anna died her body had been cremated, meaning there was no possibility of obtaining any DNA from the remains. Fortuitously however, a number of years before she died, Anna had had biopsy samples removed from her bowel, and these were stored at a hospital in Charlottesville. If permission could be gained for the team to use the samples, they could be used as a source of DNA.

But gaining permission proved no easy matter. Legally, because of the issue of patient confidentiality, the hospital could not just release the samples. It was necessary for Schweitzer to gain official permission. This proved no easy matter, largely because the Russian Nobility Association was vehemently opposed to the idea. If the tests proved Anna was genuine, there could be consequences for the association. There were also a number of other interested parties, meaning the approval process was quite complex, even before the actual DNA analysis could begin. Many people had an interest in the results, one way or another.

It was so difficult to gain permission to use Anna's biopsy samples that Schweitzer had no choice but to bring a lawsuit. Eventually, after months of effort, Schweitzer obtained permission to have access to the samples, and the research could get underway.

Peter Gill flew from England to Charlottesville to collect the samples he needed. Back in his laboratory, he and his team extracted Anna's DNA. It was now a fairly simple case of compare and contrast. DNA sequences from Anna's biopsy samples were compared with the sequences extracted from the Romanov family. They did not match. There was only one conclusion that the research team could reach—Anna Anderson could not possibly have been Anastasia.

The results of the work were soon confirmed by a second, independent group, which had managed to locate a different sample containing Anna's DNA. Amateur historian Susan Burkhart had found a lock of Anna's hair when sorting through her estate. Mark Stoneking from Pennsylvania State University and his colleagues extracted DNA from the hair, and compared it with Gill's results:

they were identical. Both samples of Anna's DNA were completely different to the Romanov DNA. There was now little doubt that Anna Anderson could not possibly have been Anastasia.

Schweitzer and his wife were naturally very disappointed, as they had always believed the DNA would prove that Anna was genuine. While he had no dispute with the scientists or their work, Schweitzer had trouble accepting the results. The way Anna was—her personality, her behaviour, everything about her—had indicated to him that she was who she claimed she was. He believed it was possible that the wrong biopsy tissue sample had been used in the analysis. It was not Anna's, he thought, but someone else's, and of course, if this were the case, the DNA would not match that of the Romanovs. While this is indeed a very remote possibility, the majority of scientists do not believe this was the case, and accept that the results of the DNA work genuinely prove that Anna Anderson was not Anastasia.

So who was Anna Anderson?

If Anna Anderson was not Anastasia, then who was she? How had she come to be found soaking wet in a river in Berlin? And why did she never reveal her true identity?

Some of those who had not believed that Anna was Anastasia had claimed she was Franziska Schanzkowska, a factory and farm worker who had gone missing in Berlin in 1920. Franziska was born around 1896 in the north of Germany, near Poland, a region called Pomerania. While working in a munitions factory in Berlin in the First World War, she was injured in a terrible explosion. One of her fellow workers died in this explosion, which left Franziska extremely upset and depressed. Perhaps wanting to forget this traumatic experience, she disappeared.

Franziska's siblings were called upon repeatedly to meet with Anna and to give their opinion as to whether she was their long-lost sister. Franziska's brother Felix had always been ambiguous about whether he thought Anna really was his sister, while Franziska's sister Gertrude said she recognised Anna as her sister, but refused to sign an affidavit to that effect. The possibility that Anna was Franziska had never been ruled out, but neither had it ever been proved.

After Anna's DNA was found not to match that of the Romanov family, Gill and his colleagues decided to carry out a little more work to establish whether Anna was really Franziska.

The scientists compared Anna's DNA with a sample donated by Carl Maucher, a great-nephew of Franziska Schanzkowska. The DNA was such a good match that Gill and his team concluded it was very likely that Anna Anderson and Franziska Schanzkowska were one and the same person.

Had Franziska, a factory worker with an uncanny resemblance to the youngest daughter of the Russian royal family, been planning all along to assume the identity of Anastasia? More to the point, did she believe it herself? According to John Klier, author of an excellent book describing the Anastasia mystery: yes, it appears she did.

Anna Anderson was not the only person to claim to be Anastasia or one of her siblings, although she was the most famous of the Anastasia pretenders. In fact, the DNA work on the Russian royal family has not stopped more Anastasia pretenders from trying to claim the Tsar's reputed fortunes. As I was writing this chapter, a news article appeared announcing that a 100-year-old Georgian woman was claiming to be Anastasia and was about to begin proceedings to claim the Tsar's fortune. There have also been, and still are, many claims from people saying they are the other children—Alexei, Olga, Tatiana—and even their descendants.

While the bodies of Anastasia and Alexei have not been found, this is not necessarily proof that they did not die at the same time as the rest of their family: they may not have been buried in the same spot. It's generally thought that even if Alexei didn't die then, he wouldn't have survived for long because of his haemophilia. Perhaps the true fate of Anastasia will never be known.

Ancient DNA has finally solved the long-running riddle of Anna Anderson, 80 years after she first appeared claiming to be Anastasia. Could other mysteries of identity be solved in a similar way? In many older cases, probably not, as a tissue sample of the mystery individual must be available to extract DNA from. There must also be a living relative of the person they are believed to be, so the two sources of DNA can be compared. Often, one or other of these conditions cannot be met, meaning that some mysteries will probably never be solved.

Solving a Titanic mystery

There is one mystery of identity, however, in which ancient DNA has played a vital role.

On 15 April 1912, the passenger ship *Titanic* began its maiden voyage from Southampton to New York. In a legendary series of events, the ship struck an iceberg about 1100 km off the Canadian coast shortly before midnight, and sank at 2.20 a.m.

The ship *Carpathia*, which was 47 nautical miles away, reached the scene three and a half hours later, saving 712 people and taking them to safety in New York. The 1496 remaining passengers were not so lucky.

In the days that followed, a grim call went out to Halifax, Nova Scotia, for ships to recover bodies. Equipped with coffins, ice, embalming fluid and undertakers, four ships were dispatched: the

Mackay-Bennett, the *Minia*, the *Montmagny* and the *Algerine*. So numerous were the dead that the crews could not cope, and a number of bodies were buried at sea. Others were taken back to Halifax for burial and possible identification.

A small child was recovered six days after the disaster by the crew of the *Mackay-Bennett*. The crew thought the boy was about two years old, and the sight of him brought the sailors to tears. They decided they would take responsibility for burying him if his body was not claimed.

The boy was not identified, so he was buried by the crew of the salvage ship in Halifax's Fairview Lawn Cemetery amongst 150 others from the disaster. His headstone bore the simple inscription: 'Erected to the Memory of An Unknown Child'.

Over the decades that followed many guesses were made about the child's identity. The strongest theory—believed by the coroner at the time of the disaster and by many people since—was that the child was 2-year-old Gösta Leonard Pålsson, from Sweden. However, this could not be proven and, as there had been a number of young children on board the ship, there were several other possibilities for the child's identity.

For 86 years, the child's identity remained a mystery, until, in the late summer of 1998, DNA expert Ryan Parr from Genesis Genomics Inc. and Lakehead University in Ontario, and co-investigator Alan Ruffman from Geomarine Associates in Halifax, obtained permission to exhume the remains of the child and two others from the cemetery whose identities were also unknown. Their aim was to try to solve the puzzle by extracting the victims' DNA and comparing it with DNA from families who had lost relatives in the *Titanic* disaster.

On 18 May 2001, the child's grave was opened, and it soon became apparent that analysis of his DNA was not going to be easy. The soil was slightly acidic and damp, and the child's remains had virtually disappeared. The researchers managed to remove a tiny six grams of bone and three teeth from the boy, but the other two bodies had completely decomposed, leaving no hope of identifying them.

Led by a team at Lakehead University's Paleo-DNA Laboratory, the researchers began the difficult task of extracting the DNA from the bones. The poor condition of the remains made this complicated, but after some effort they did succeed in obtaining DNA sequences of fragments of the boy's DNA, and replicating the results in independent laboratories—making them confident the DNA sequences were authentic. The researchers compared the child's DNA with that of a maternal relative of the Pålsson family—and it did not match. The child could not possibly be Gösta Leonard Pålsson after all.

The search was on to find the child's true identity. A thorough analysis of the teeth revealed that the child was very young indeed, possibly less than a year old. The *Titanic*'s passenger lists revealed five male children around this age in addition to Gösta Leonard Pålsson. Could the unknown child be one of them?

Because of the ever-present potential for contamination, to be absolutely sure they had the correct DNA sequences from the child the team extracted DNA from his teeth, and repeated the bone DNA analysis. Extensive genealogical work was carried out to locate relatives of all of the remaining candidate children, in order to obtain DNA samples. The families involved were generous with providing samples, and the researchers were able to acquire the DNA they needed.

The unknown child's DNA was compared with that of all the candidate families, and, finally, the answer to the mystery was revealed. The DNA was a perfect match to Magda Schiefer from

Finland, the great niece of the mother of the 13-month-old boy, Eino Viljami Panula. Eino, from Ylihärmä in Finland, had travelled in third class on the *Titanic* with his mother and four brothers, to join their father in Pennsylvania. None survived the journey. 'The unknown child is now a known child, identified and returned to his family,' said Parr.

A T C G

There is one other, much older mystery of identity in which DNA analysis has been extremely useful, involving the identity of another child. Louis XVI of France and his wife Marie-Antoinette were executed in the bloody events of the French Revolution and their son, Louis-Charles, was recorded as having died in prison soon afterwards. A number of years later, a man named Carl Naundorff, claiming to be none other than Louis-Charles, appeared. Whether he was genuine could not be proved—until now.

7

THE HEART OF THE MATTER

What became of Louis XVII of France?

'How ugly everything is here,' Louis-Charles complained to his mother as he surveyed his surroundings. His new home was quite a shock for the 4-year-old prince, who, as heir to the French throne, had spent all of his young life in the lap of luxury. The dilapidated and disused palace of the Tuileries had not been lived in by a royal family for generations, but Louis-Charles, like his parents, had little choice in the matter. It was 1789 in Paris, and the French Revolution had just begun.

Louis-Charles was born on 27 March 1785 in the sumptuous palace of Versailles, the home of King Louis XVI of France and his wife, Marie-Antoinette. He was the King and Queen's second son, and their third child. Louis-Charles, like his father, was descended from a long line of French monarchs, the Bourbons, who had ruled France for centuries.

When Louis-Charles was born, the royal family enjoyed a privileged existence, surrounded by courtiers and servants. They believed they had the loyal support of the French people—after all, Louis XVI had been hailed enthusiastically by his subjects when he ascended the throne as a young man. Outwardly, the royal family enjoyed a peaceful and secure life. But under the surface, a time bomb of discontent was brewing. Amongst the people of France, all was not as it seemed.

The country was in truth in a dire state, both financially and socially. Population increases in recent years had led to higher grain prices, lower wages and reduced standards of living. Harvests had been disappointingly meagre, which drove up the price of bread, the nation's staple food, even further. To compound matters even further, war in the preceding years had left the state in considerable debt. The problems were worst amongst those who were less well off, and hundreds of thousands of ordinary people were desperately poor and hungry.

King Louis XVI was by all accounts a good-natured, pleasant man, with the best of intentions for his family and his country. Sadly, though, his privileged lifestyle prevented him from fully understanding the plight of the average citizen. Poverty was something he just could not comprehend, and so, despite some genuine attempts, he was not able to find ways to ease the people's suffering in any significant way. Looking for a scapegoat, people began to blame their miserable economic situation on the extravagances of the court, and resentment towards the royal family grew stronger by the day.

The situation reached breaking point in 1789 when Louis-Charles was just four years old. On 14 July a furious crowd stormed the Bastille in Paris, the notorious prison that had long been a symbol of royal power. Prisoners were released, and the building itself was destroyed.

Over the following months, the uprising continued, until one night a ferocious and bloodthirsty mob, bearing the severed heads of loyal soldiers from the King's Guard, arrived on the royal family's doorstep in Versailles. The King and his family had no choice but to leave their home and retreat to the disused and run-down palace of the Tuileries, in Paris.

The riots continued outside the gates of the Tuileries, and the royal family feared for their lives. In desperation, they made plans to flee the country. In the dead of night, the children were woken from sleep and bundled into a carriage. Disguised as ordinary travellers, the family fled as quickly as they could through the night, towards the German border.

They maintained breakneck speed until dawn, when their horses almost collapsed from exhaustion and they were forced to stop. Unfortunately, it did not take long before they were recognised. The King and his family were forced to return to Paris to face the waiting crowds.

Far from being the deliverance they had so hoped for, this thwarted attempt at freedom only made things worse. Rumours abounded that the family had been trying to enlist the help of foreign armies to attack their own country. The crowd grew more bloodthirsty than ever. Louis XVI was being forced ever deeper into a revolution over which he had no control.

It had been a terrible year for the royal family, a year in which they were reduced from absolute monarchs leading a life of wealth and privilege to virtual prisoners in a crumbling palace. To make matters worse, the family had also suffered a personal tragedy. In that same year, the King and Queen's eldest son, the heir to the throne, died after a long and painful battle with tuberculosis. Four-year-old Louis-Charles, the new heir, was suddenly thrust into the limelight at one of the most dramatic times in the history of his country.

The following three years passed in an uneasy kind of peace for the royal family. Louis-Charles was given a tutor, and the family tried to continue life as normally as they could, all too aware that they now had no real power, and no real support.

But outside the palace, discontent was still brewing. Riots occurred periodically, and anti-royalist feeling grew ever stronger. Finally, in a dramatic uprising in August 1792, the Tuileries palace was set on fire, and the royal family was forced to seek refuge in the National Assembly.

Louis XVI and his family were taken to the tower of the Temple of Paris, a bleak and imposing building which had formerly been a medieval fortress and which now became their prison. Watched over by 25 guards, their correspondence was cut off, and they were given not enough clothing or other household essentials. All but two of their servants and attendants were sent away.

After the initial shock wore off, the royal prisoners settled again into a kind of day-to-day routine. They would eat breakfast together, and every morning at ten o'clock the King gave lessons to his son, watched over at all times by guards. Only certain subjects were permitted, however, and not mathematics, in case a secret code was contained in the numbers. Louis-Charles was allowed to play in the garden some days, but not always.

But this relatively peaceful existence was not to last for long. Outside the Temple, resentment towards the King was still growing. Completely ignoring his desperate pleas not to take action, foreign monarchs had sent their armies to invade France. Naturally, the people blamed the King. He was forced to renounce his royal title, and the watch on the family grew tighter.

As time went on, the situation grew worse. The revolutionaries had been looking for an excuse to bring the King to trial for some time, and the discovery of documents he had hidden in a secret chamber was just the excuse they needed. Louis was forcibly separated from his family and tried before the National Assembly.

Faced with the weight of the mountain of 'evidence' brought against him, most of which was fabricated, the King must have known what would come next. On Christmas Day of that year he made his will, forgiving all who had ever harmed him.

Inevitably, he was sentenced to death. King Louis XVI, who until a few years before had been an absolute monarch, free to rule his country in any way he saw fit, was executed by guillotine on 21 January 1793.

As it happened, the call for his death was decided by just one vote, with 361 for to 360 against. The deciding vote was cast by his cousin, the Duc d'Orleans. His own family had betrayed him.

Louis's brother, the Comte de Provence, immediately proclaimed Louis-Charles, still only eight years old, as King Louis XVII. Provence then appointed himself Regent of France until Louis-Charles came of age, and appointed his younger brother, the Comte d'Artois, Lieutenant-General. England, Russia and the United States, together with the few royalists left in France, supported the young Louis-Charles as King and increased their campaign against the French revolutionaries.

The leaders of the revolution began to get very nervous. The threats posed by the advancing foreign armies and the declarations of the Comte de Provence were added to by the discovery of a prophecy, supposedly originating from Nostradamus's father, indicating that power would be restored to the royal family. The result was that Louis-Charles, now seen as a threat, was torn struggling from his mother's arms and handed over to Antoine Simon and his wife Marie, both fervent revolutionaries.

This traumatic turn of events marked the beginning of another tragic chapter in the sad young life of Louis-Charles. In a few short years he had seen his brother die and his father beheaded, and now he was wrenched away from his mother and sister, to be locked away in a small room. Nor were things about to improve for the young king.

From that point on, Louis-Charles was not permitted to see his sister or any other friends or family members. He was also kept away from the public eye, so exactly what happened to him during this time is something of a mystery. It is known that the idea behind his confinement was to 're-educate' him, to keep him separated from royalists and his family and manipulate him into thinking along the same lines as the revolutionaries. He was forced to wear revolutionary clothing, sing revolutionary songs and denounce his family. There are rumours that Simon may have mistreated Louis-Charles, and although this has never been proven, it must have been a pretty miserable life for the young boy.

On 1 August 1793 his mother, Marie-Antoinette, was also put on trial. Evidence against her was almost certainly fabricated—even her own children were forced to sign accusations against her. The trial ended, inevitably, with her execution.

Louis-Charles remained imprisoned in the Temple for several more years, separated from everyone and everything, with a succession of watchful guardians assigned to his care. His living conditions became even worse when he was locked in a small room with no light, little food and no toilet facilities. There he remained, day after day, becoming weaker and more desperate. His sister appealed to her captors to be allowed to look after her brother herself, but her pleas fell on deaf ears.

In early 1795, it became apparent that Louis-Charles was gravely ill. A physician was called to attend the boy, and his room was at last cleaned, but it was too late. On 8 June 1795, aged ten, Louis-Charles, the uncrowned King of France, died. He had succumbed to tuberculosis, undoubtedly as a result of his appalling living conditions.

A post-mortem was carried out the next day. During the procedure, Dr Philippe-Jean Pelletan removed the dead child's heart as a grotesque souvenir, pocketed it, and took it home with him where he pickled it in distilled wine alcohol. His actions, though macabre, were not entirely unusual, as the collection of body parts of famous people as mementoes was quite popular at the time. Little did he know just how fortuitous this action would prove to be.

Louis-Charles's death was announced, and he was buried in a mass grave in the churchyard of Sainte-Marguerite. The child's life had ended—but the mystery of Louis XVII had only just begun.

Almost immediately, people began to whisper that Louis-Charles was not dead—that the prince had been exchanged at some earlier time for another child, who had died in his place. The real Louis XVII was now hidden somewhere, most likely away from France. There was absolutely no proof that the child who died was the genuine Louis-Charles. He had been hidden away, locked in the Temple for quite some time, and although he had been allowed a few visitors just before his death, no one could guarantee that he had not, by that stage, already been exchanged for another child.

Even the warders who had guarded him could neither confirm nor deny whether the child who died was the real Louis XVII. Even Pelletan, the physician who performed the post-mortem, had no idea if the body he was examining was that of the real Louis-Charles. To add to the mystery, no one who knew him as a young child, not even his sister, had been permitted to identify the body.

Adding fuel to the rumours, in 1811 a 78-year-old woman came forward, claiming to be Marie Simon, the wife of Antoine Simon, Louis-Charles's first warder when he was separated from his parents in the Temple. Now incarcerated in the interestingly named Hospital for Incurables, Madame Simon told how Louis-Charles was smuggled out of the Temple in a washing basket, and another child smuggled in to take his place. Louis-Charles was still alive, she swore. In fact, she said, he had even visited her in hospital some eleven years earlier. Marie Simon was never allowed to state her claims in open court. Her story was never verified—but it added to the rumours nonetheless.

After the restoration of the monarchy in the early 1800s, King Louis XVIII decided to resolve the mystery once and for all. He ordered a search for the remains of Louis XVI, Marie-Antoinette and Louis-Charles. But the remains of Louis-Charles were nowhere to be found.

Had royalist supporters smuggled Louis-Charles out of the Temple, and replaced him with another child? Was Louis-Charles still alive? In the absence of any compelling evidence either way, people were free to believe what they wished. And believe they did. Years and then decades passed, and still the rumours did not die. But if Louis XVII were still alive, where was he?

Was Karl Wilhelm Naundorff Louis XVII?

One spring day in 1833 a down-at-heel traveller arrived in Paris. He had just come from Germany, where he had been living in recent years. His passport was in the name of Karl Wilhelm Naundorff, watchmaker by trade. But this was no ordinary down-on-his-luck itinerant with a poor grasp of French. Karl Naundorff had a secret that he was ready to share. He wanted the whole world to know that his real name was Louis-Charles, King Louis XVII of France, and he was determined to prove it.

In the 35 years since Louis-Charles's death was announced, there had been quite a number of other men claiming to be the young heir. All were soon revealed as the impostors they were. Unsurprisingly, most people initially thought that Naundorff, too, was a fraud. But this time it was not so easy to dismiss his claims.

For a start, Naundorff bore a striking physical resemblance to Louis-Charles. He also had a number of specific characteristics that Louis-Charles had been known to have, such as a triangular vaccination mark, a pigeon-shaped mole on his thigh, a scar on his top lip, and prominent teeth.

Naundorff also had vivid recollections of the childhood of Louis XVII. He knew intimate details that only those who were there at the time could possibly have known. He was also able to describe in detail his escape from the Temple of Paris.

These recollections, the account of his escape and his physical similarities to Louis-Charles soon had the majority of the public convinced that he was who he claimed to be. Even more interestingly, Naundorff was recognised by some who had actually known the real Louis-Charles. Shortly after his arrival in France, Naundorff contacted several surviving members of Louis XVI's court, including Monsieur de Joly, the last Minister of Justice, Madame de Rambaud, Louis-Charles's own governess, and Monsieur and Madame Marco de St-Hilaire—the dead king's chamberlain and lady-in-waiting to the king's Aunt Adelaide. All four recognised him.

Madame de Rambaud in particular was easily and utterly convinced. She said in a written statement that in her conversations with Naundorff they had 'exchanged recollections which alone would have been incontrovertible proof for me that the Prince is really what he pretends to be, the orphan of the Temple'. Marco de St-Hilaire also made a written declaration—he too was convinced that Naundorff was genuine. Monsieur de Joly, although at first sceptical, was also soon won over.

But if Naundorff really was Louis XVII, where had he been for the last 30 years or so? And how had he escaped from the Temple, where he had been so closely watched? In response to the inevitable questions, Naundorff began to reveal his life story.

To escape from the Temple, Naundorff explained, he had been drugged with opium, and carried out in a coffin. In preparation for his escape, another similar-looking child had been smuggled into the Temple some time earlier, and it was this child who had died and been buried as Louis XVII.

Naundorff escaped from France, he said, by pretending to be the son of a Swiss woman. For many years, he continued to move across Europe, and was periodically imprisoned. Finally, in 1810, he found himself in Berlin, and established himself in the Schutzenstrasse as a watchmaker.

Soon afterwards he was asked by the authorities for a passport and, when he could not produce one, was given one in the name of Karl Wilhelm Naundorff, watchmaker, born in Weimar. He had used this identity to reside in several towns in Prussia, before arriving in France in 1833.

Was this a carefully fabricated lie by a clever con man, or could it be the incredible true-life story of the unfortunate young heir to the French throne? Could Karl Naundorff, by now middle-aged, really be the long-lost Louis-Charles?

Naturally, Naundorff's background was soon investigated, and it was certainly true that Karl Naundorff was the name given to him after his arrival in Berlin. But no record could be found of his life before this, despite repeated attempts. If he was not Louis XVII, his real origin was a complete mystery.

There was one person who, in theory, could have verified or refuted Naundorff's claims. Louis-Charles's sister, the Duchess d'Angouleme, had eventually been released from the Temple and was still alive at the time Naundorff made his appearance.

Naundorff had actually tried to contact her some years earlier when he was still living in Prussia. 'My dearly beloved sister,' he wrote, 'I tell you, I am alive; it is I, your real brother. Ask me to prove it. I pledge myself to do so.' But his pleas fell on deaf ears, as the Duchess steadfastly refused to answer this, or any of his subsequent letters. Although she had not been permitted to see the dead child in the Temple, she was certain that her brother had died, and that Naundorff was a fraud.

Now that Naundorff was in France, he made a renewed attempt to contact the Duchess, both in person, and through his most important supporters. Madame de Rambaud wrote to the Duchess, to 'assure you of the existence of your illustrious brother. My eyes have seen him and I have recognised him'. But again, the Duchess would not respond. Bremond, Louis XVI's private secretary, also wrote, assuring her of his belief in Naundorff's authenticity: 'As a servant of your illustrious father I have recognised in Karl Wilhelm Naundorff the orphan of the Temple . . .' Once again, there was no response.

Unable to win the attention of the Duchess, in 1836 Naundorff began a lawsuit with the aim of being officially recognised as Louis XVII. Having had no success through approaching Louis-Charles's relatives informally, he had summonses sent to both King Charles X and the Duchess d'Angouleme, requiring them to appear before a court to answer his claims.

Before the summonses could be delivered, however, Naundorff was arrested. The collection of more than 200 documents which he intended to use to present his case was confiscated, and he was imprisoned without a warrant.

Despite his lawyers' best efforts to have him released and the trial re-activated, Naundorff was expelled to England. Obviously he was seen as a threat by the French authorities, whether he was Louis XVII or not.

Naundorff had no choice but to settle in England, where he continued his campaign to be recognised officially as Louis XVII. He wrote a book of memoirs, which, unsurprisingly, was not allowed to enter France. In general, his life in England was fairly happy, as he had a lot of support and loyal followers there. But it was not without incident.

One foggy winter evening in 1838, Naundorff felt the call of nature, and ventured down to the bottom of his garden, where the lavatory was located. A few seconds later, the servants in the house heard shots and cries. They rushed outside and found Naundorff bleeding on the ground.

Fortunately, Naundorff survived the murderous attack, a miracle considering his assailant had aimed at his chest and the bullet had been deflected only by his arm, which he had held up to protect himself. The assailant was soon captured, and identified as one Mr Roussel, a French citizen. The attack was almost certainly politically motivated.

This was not the only dramatic episode during Naundorff's time in England. In his spare time, he enjoyed the rather eccentric pastime of experimenting with bombs, guns and explosives. For this activity, he had a workshop especially set up at his home.

One day, he returned from town and went to his workshop. It must have been quite a shock when, without warning, a corner of the workshop burst into flames. The fire quickly spread, and soon the whole room was alight.

Naundorff, trapped in the room, was most alarmed by his predicament, especially as he knew there was a container of explosives near the window. There was no other option—he dived into the flames to get the explosives so he could throw them out the window. As he grasped the box, it exploded, badly burning his hands and his face. Again, this was a politically motivated attack—someone had broken into the workshop in his absence and set the fire.

Despite these dramas, Naundorff was generally very popular in England, where he was always referred to as the Duke of Normandy, the title of Louis XVII. When his son was born in 1840, the boy's birth certificate identified him as 'Adalbert, Prince of France, son of His Royal Highness le Duc de Normandie'. But recognition still eluded him in France. In 1840 Naundorff again tried to get his case heard before a French court, but it was again rejected.

In 1845 Naundorff moved to Holland, hoping to sell his weapons inventions to the Dutch Government. He had first offered them to the French Government which had, not surprisingly, shown no interest. At first the Dutch Government was suspicious, but after some negotiation, they did agree to buy his inventions.

At last, it seemed, he would be financially secure. But his happiness was short lived, as he fell gravely ill and died on 10 August 1845. He was buried in Delft in the Netherlands. In an extraordinary move by the Dutch Government, his death certificate was made out in Louis XVII's name, although his claims had never been proven, nor his true origins ever established.

Naundorff had never given up his insistence that he was Louis XVII, even to his own wife and children. On his headstone was placed an inscription which translates as:

> *Here lies Louis XVII, Charles Louis, Duc de Normandie, King of France and Navarre, Born at Versailles on March 27, 1785, Died at Delft on August 10, 1845.*

Naundorff's death did not put an end to the mystery. His family continued to seek recognition as members of the royal family and, in the public domain, the mystery of his identity was still hotly debated. Well into the nineteenth century, what had really happened to Louis-Charles was still a complete puzzle.

Over the next 100 years, numerous historians closely examined the existing evidence. Each drew their own conclusions, some believing that Louis-Charles had escaped and Naundorff was who he claimed to be, others that Louis-Charles had died, lonely and abandoned, in the Temple. But no definitive proof could be discovered. It was clear that new evidence would have to come to light if the mystery were ever to be laid to rest.

One day in 1992, more than 200 years after the bloodthirsty days of the French Revolution, Professor Jean-Jacques Cassiman was contacted at his office at the University of Leuven, in Belgium, by Dutch historian Hans Petrie. Petrie had a proposition for Cassiman, a proposition that he believed could at last solve the Naundorff mystery.

The key to the idea was an interesting discovery that Petrie had made some time earlier. While researching the mystery, Petrie had uncovered a theory that Naundorff might have died from arsenic poisoning. To verify the truth, Naundorff's coffin had been opened in 1950 and his right humerus and a lock of hair removed for testing. This investigation revealed no traces of arsenic, but the bone was now stored in the archives of the Dutch Forensic Laboratory in Rijswijk, and the lock of hair in the mayor's office in Delft.

Petrie wondered whether these remains might still contain traces of Naundorff's DNA, and if they did, whether it could be compared with DNA from known relatives of Louis-Charles. If the DNA matched, this would be definitive proof that Naundorff's claims were true. If the DNA did not match, Naundorff would be revealed as an impostor.

Experts were needed to carry out the specialist DNA work, and Cassiman and his research team were the perfect choice. They had a wide range of experience in using DNA to solve paternity disputes, and had also been involved in extracting DNA from ancient bones uncovered by archaeologists from sites in Turkey.

Cassiman had no hesitation in admitting that his interest in the case was purely scientific. Would it be possible to extract DNA from Naundorff's remains and, if it was, would it be possible to prove whether Naundorff was who he claimed to be? But as for the possibility of unravelling one of the most intriguing real-life mystery stories of recent history, Cassiman was quite frank. 'The history itself—I couldn't care less,' he said. 'It was a challenge—a technical challenge.'

He faced an initial setback when, in spite of 62 painstaking attempts, the group was unable to extract any DNA from Naundorff's hair. They had more success with the humerus, and were able to extract a decent sample of DNA from the bone. Why did they have such difficulty extracting DNA from the hair sample? 'It is hard to identify the reason,' said Cassiman. 'I presume that the way [the hair] was stored affected the DNA. Alternatively, the hair may have been treated by a product that degraded the DNA.'

Armed with a sample of Naundorff's DNA, the next stage was to obtain DNA samples from close relatives of Louis XVII. This would prove to be the biggest challenge of the whole project. After a huge effort, no suitable sources had been found, and the situation was beginning to look a little grim.

Then the researchers made a wonderful discovery. Louis-Charles's mother, Marie-Antoinette, was one of 16 brothers and sisters. Her mother, Maria-Theresa, had kept a chain of rosary beads with 16 gold medallions attached to it, each medallion containing a lock of one of her children's hair.

Marie-Antoinette's eldest sister, Maria-Anna, had inherited the rosary when their mother died. Maria-Anna spent the last years of her life in a convent in Klagenfurt, Austria, and her possessions, including the rosary, had been left to the monastic order.

The hair in the rosary was of course that of Louis-Charles's mother, aunts and uncles, and therefore a perfect potential source of DNA for comparison with Naundorff's DNA. It was the breakthrough the researchers had been looking for. Because of the historical value of the rosary, Cassiman's team was able to gain permission to use hair samples from only two of the medallions: those of Louis-Charles's aunts Johanna-Gabriela and Maria-Josepha. The medallions were carefully opened, and a few hairs taken from each.

At about the same time, the team was able to make contact with two living relatives of Louis-Charles, Queen Anna of Romania and her brother, André de Bourbon Parme, both of whom agreed to provide samples. Finally, in a stroke of luck, the team managed to locate two samples of hair from Marie-Antoinette herself.

Back at the laboratory, the researchers painstakingly extracted DNA from all the samples. It was time for the moment of truth. Would Naundorff's DNA match the DNA of Louis-Charles's relatives? With the utmost care, Cassiman and his team compared the sets of DNA—and found that Naundorff's DNA was completely different.

There was only one logical conclusion to be drawn—Naundorff could not possibly be the son of Marie-Antoinette, and thus could not possibly be Louis-Charles. Karl Wilhelm Naundorff, the watchmaker who had convinced so many that he was the son of royalty, was an impostor.

If Naundorff was not Louis-Charles, what had happened to the young king? Did Louis-Charles die in the Temple of Paris? Or was he substituted for someone else altogether? Could the DNA that had revealed Naundorff's secret shed light on this part of the mystery as well?

Cassiman knew that if some genuine DNA from the boy who died could be located, it could be compared with that from Louis-Charles's relatives. If it matched, it was indeed Louis-Charles who had died. If it did not, a substitution had taken place.

The researchers knew that the dead boy's heart had been removed and pocketed by the surgeon who performed his autopsy. If the heart still existed, the DNA it contained could perhaps solve the mystery.

Locating the heart, however, was no easy matter. 'We knew the heart was somewhere, but we didn't know where,' said Cassiman. Luckily, many historians were also eager to see the mystery of Louis XVII solved. One of them, Philippe Delorme, contacted Cassiman out of the blue and told him he had found out where the heart was kept.

A number of years after performing the autopsy, Pelletan decided it was the right time to return the heart to the French royal family. But when he went to retrieve the heart, he realised it had been stolen. Fortunately, it was soon returned to him by a relative of the thief—one of his former assistants had taken it.

Pelletan offered the heart to Louis XVIII, Louis-Charles's uncle. Although he was quite insistent that the king take the relic, Louis XVIII refused to accept it. Pelletan then offered it to Louis-Charles's sister, the Duchess d'Angouleme, who also refused it.

By then, in 1828, Pelletan was 81 years old, and dying. In desperation, he finally gave the heart to Monsieur De Quelin, the Archbishop of Paris.

The heart was hidden in the archbishop's palace, but in 1830 the palace was plundered, and virtually all its contents stolen or destroyed. Miraculously, Pelletan's son found the heart in the remains of the palace: it had been trampled into the ground and was covered in dirt, but was still intact.

Pelletan's son put the precious relic in a crystal urn, in which it is still kept, and held it safely at his house for the next 50 years. Upon his death in 1879, the family again tried to return the heart to the royal family, but this effort, too, came to nothing. The heart was passed on to Eduard Dumont, a relative of Pelletan's son's wife.

Dumont asked the Spanish Bourbons if they would like the relic. They accepted and, in 1895, Don Carlos de Bourbon, head of the Spanish branch of the Bourbon family and grand-nephew of Louis-Charles, placed the heart in the chapel of the Château de Frohsdorf in Austria, where it stayed for the next 50 years.

The chapel was looted during the Second World War, and the heart disappeared once more. Extraordinarily, however, it had escaped injury yet again, as two of Bourbon's grand-daughters had rescued it. They offered it to the Duc de Bauffremont, president of the Memorial of Saint-Denis in Paris in 1975. Here it was to reach its final resting place.

The Cathedral of Saint-Denis houses a crypt, where the remains of many members of French royalty are kept. It also houses a number of urns containing preserved organs from the kings of France, including hearts, and the heart of the boy who died in the Temple was placed there.

The Duc de Bauffremont had the final say over whether the heart could be used in Cassiman's research. He decided to allow the DNA analysis to go ahead, hoping that the truth would, at last, be revealed.

Under the avid eyes of a crowd of spectators, the heart was removed from the crypt, and a blessing ceremony held. It was taken, still in its crystal urn, to a nearby laboratory, where two samples

The crystal urn containing the heart of the young Louis XVII on display in Saint Germain l'Auxerrois church in Paris shortly before the heart was finally entombed. (Reuters / Picture Media / Victor Tonelli)

were taken from it and sent to two independent laboratories. One was Cassiman's laboratory in Belgium, the other a laboratory in Germany that Cassiman had asked to assist with the research.

Both laboratories managed to extract DNA from the samples. When compared with the DNA of Louis XVII's relatives that had been used in the Naundorff study, a perfect match resulted. The heart could be that of none other than Louis XVII.

The news that the official version of events was correct and that Louis XVII had died a prisoner in the Temple of Paris had quite an impact, especially on Naundorff's descendants. They had initially been very excited about the research being carried out because they genuinely believed that Naundorff was who he claimed to be. The result left them much less excited; in fact, they refused to believe it.

Others, too, were sceptical. 'In this whole story there are two groups, the believers and the non-believers,' said Cassiman. Although most people were willing to accept the results of the DNA work, a number of Naundorff's most loyal supporters still clung to the hope that the DNA evidence was wrong and that Naundorff was indeed who he had claimed to be. 'They're never going to give up. That's clear,' Cassiman explained. 'It's a belief. It's not a science, it's a belief.'

Cassiman, however, does concede that there is a remote possibility, just as there was in the Anna Anderson case, that a mix-up could have occurred and the bone from which the DNA was extracted was not Naundorff's at all. 'I can't prove that,' said Cassiman, 'except that I have all the documents proving how this bone travelled to the lab where we recovered it from.' The authenticity of the heart of Louis-Charles has also been called into question by those who do not believe he died in the Temple. It has been suggested that it was Louis-Charles's brother's heart that was preserved, as he had died only a few years earlier. 'Of course, I cannot prove that is not correct with what we did,' said Cassiman, 'and that's why I say the historians are responsible for the proof that the heart is indeed that of [Louis-Charles].'

'The only thing that we can prove,' he said, 'is that it is the heart of a child—that's definite—and that its DNA is identical to that of the Hapsburgs and of his presumed mother, Marie-Antoinette. But that's as far as we can go. The historians have good arguments, they tell me, to say that this is the heart of the child that died in the Temple. But that's not something I can prove.' When asked whether he personally believed that the heart really was that of Louis XVII, Cassiman conceded: 'Yes, it is very unlikely that it is someone else.'

Cassiman received no official comment from the French Government about his results. 'It's very tricky, because we're talking about the heir to the throne—the French throne—and there are two factions in that family, who fight continuously, the Spanish branch and the French branch, and . . . well, I don't want to get involved in that, it's their problem,' he said with a wry chuckle.

Ancient DNA has now solved one of the most intriguing mysteries of recent history. The short and tragic life of Louis-Charles, the young King of France, *did* end in the Temple of Paris, most likely as a result of tuberculosis. Karl Naundorff, who managed to convince so many that he was Louis XVII, was revealed as a fraud.

Despite the tremendous success of the research, a number of questions remain unanswered. Who was Naundorff? Where did he come from? Did he really believe that he was Louis XVII, or was he just a fantastic liar? And as for Louis-Charles—did he really die from tuberculosis? Could he have died from neglect instead, or even been murdered? The answers to some of these questions may never be known.

CONCLUSION
What next for ancient DNA research?

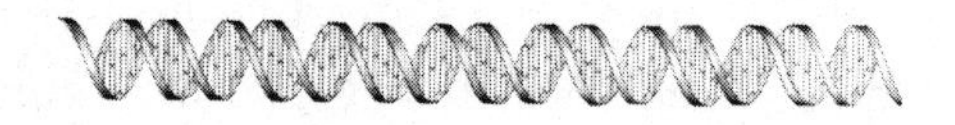

The diverse field of ancient DNA research has achieved a great deal in the relatively short time that the methods for it have existed, and even more exciting possibilities are beckoning for the future of the technology. Ancient DNA research is, by its very nature, an extremely diverse field with projects involving a wide variety of areas of science. The directions it could take in the years to come are limited only by the availability of suitable samples.

It has now been established that DNA can be found in a wide range of ancient tissues, which means that there is an abundance of avenues for future research. 'We're still surprised where we find it,' says Alan Cooper. 'In terms of tissue samples, we have bone, teeth, soft tissues, muscles . . . faeces work very well indeed—it's a very active area of research—and hair.' In theory, any biological sample less than about 100 000 years old is a potential candidate for ancient DNA research if it has been well-enough preserved.

The future possibilities for ancient DNA work, therefore, are almost endless.

Infectious diseases

Many of the avenues of investigation discussed in this book still present rich opportunities for further research, one example being the history of infectious disease. As well as the work involving tuberculosis in the Americas, the Black Death and the 1918 flu, ancient DNA has been extracted from a range of other bacteria, viruses and parasites that have caused historical outbreaks of illness: the bacterium that causes leprosy, *Mycobacterium leprae*; the parasite that causes malaria, *Plasmodium falciparium*; the parasite that causes Chagas disease, *Trypanosoma cruzi*; the bacterium that causes syphilis, *Treponema pallidum*; the gut bacterium *Escherichia coli* (*E. coli*), and the bacterium that causes diphtheria, *Corynebacterium diphtheriae*.

Ancient DNA from most of these diseases has been found in just a few samples, showing that it is possible to extract DNA from a range of disease-causing organisms residing in ancient human tissue. In the future, more research in this area could give us a detailed picture of the evolution of a range of human diseases, as well as how historical communities were affected by disease.

More can also be learned about the three diseases discussed in this book. Tuberculosis has proven to be a particularly suitable disease for ancient DNA analysis, as the bacterium that causes it is preserved in ancient tissue particularly well. Future studies could examine issues such as how frequent tuberculosis was in different human populations in the past and how much mortality it caused.

More research also needs to be carried out into the issue of whether the Black Death was caused by the plague, and there is still more that could be learned about the origins and deadliness of the 1918 flu.

Extinct animals

The future of ancient DNA research will also involve more investigation into the DNA of extinct animals. There is still the possibility that one or more extinct species may be cloned one day, and this continues to be a small, but active, area of research. However, most of the ancient DNA research into extinct animals is now focused on learning more about the ecology of ancient animal communities.

Ancient DNA from animal remains buried in permafrost is a particularly promising field of study. Cooper and his team regularly extract DNA from animal remains buried in Alaska and Siberia—an activity which, Cooper jokes, is fabulous: 'as a threat to students who aren't working hard enough, to send them to Siberia is actually quite a good one'.

Cooper and his team are using this DNA to investigate the effects that climate change during the Ice Age had on the size, range and genetic diversity of populations of Ice Age animals. This research could be informative for predicting the effects of global warming on present-day species and populations, but is also giving more clues in the debate about what caused the Ice Age megafauna to become extinct. Before he began this work, Cooper shared the view that humans very likely had a major role in killing the megafauna. He now believes, however, that Ice Age climatic conditions did the real damage. His work indicates that the few populations of megafauna left after the height of the Ice Age were in a weakened state, which made it very easy for humans to cause them to become extinct. 'They were just like a dead man walking,' Cooper says. If humans had not arrived, some of the weakened populations of megafauna might have been able to survive, but this is by no means certain.

DNA from dirt

While they were working in Siberia, Cooper and his team discovered something which they believe may lead to another exciting line of ancient DNA research. Part of their work involved drilling holes through the permafrost to take core samples. Siberian permafrost is about 2–3 million years old, and Cooper wondered if the core samples might contain DNA left behind from the multitude of animals, plants and bacteria that had lived throughout this time. 'What the hell,' thought Cooper, 'we'll have a go and just see if we can amplify any DNA from plants or animals.'

Cooper was indeed able to extract mammoth, bison and plant DNA from the core samples and, as deeper parts of the ground date back to earlier time periods, was even able to see the DNA changing—evolving—through time. This was truly revolutionary, as it had always been thought that physical remains—bones, teeth or pieces of skin, for example—are necessary to study the DNA of ancient animals and plants. The question now is, how much more can the permafrost reveal about extinct ecosystems through the ancient DNA contained within it?

Cooper went on to try the same approach with some earth he had kept from a New Zealand cave when extracting moa bones. To everyone's amazement, this dirt also contained DNA: from moas and other extinct New Zealand birds, as well as trees no longer present in that part of New Zealand. Cooper verified that he had extracted genuine DNA from these species by matching it to modern-day samples that he knew for sure were from these animals and plants. 'We got a complete picture of [the flora and fauna of] New Zealand prior to human colonisation in that area.' At the moment, he says, 'we don't even know what we can do with this stuff [but] there's a whole genetic record of what the environment was like'. The results suggest that it might now be possible to build up even more intricate pictures of ancient ecosystems, including what species lived during which time periods and in what locations, from the DNA contained within layers of dirt devoid of any trace of visible fossils.

Ancient DNA has a bright future, with many exciting possibilities yet to come, and what the field has achieved in a few years' time will no doubt provide enough material for another whole book on the subject. One thing is for sure, the future of ancient DNA is well and truly in the past.

SOURCES

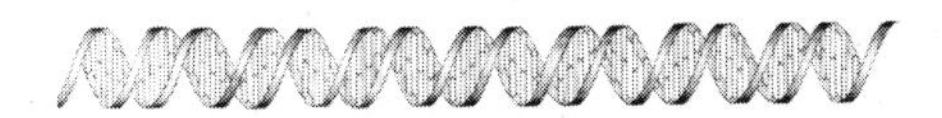

Writing this book involved coming to grips with a wide range of materials and subject matter. Because of this, I frequently found it necessary to rely on secondary sources to provide information, including textbooks and popular books on relevant topics, together with a number of relevant and reliable websites. Having these excellent and often very readable sources available made what would otherwise be an almost impossible task much easier and more enjoyable.

Together with the secondary sources, a range of primary sources of information was used which included published articles in scientific journals, and interviews and correspondence with researchers involved in the ancient DNA field. I was fortunate enough to attend a number of entertaining and interesting seminars on ancient DNA research, which also provided valuable information.

At times, I uncovered conflicting information from the various sources on a particular topic. When this occurred I endeavoured to uncover the consensus opinion. Inevitably, errors may remain because of this.

INTRODUCTION

Part of the background information on DNA was sourced from the following book and website:

Anonymous 2004, 'How to extract DNA from anything living', *Genetic Science Learning Center: University of Utah*, <http://gslc.genetics.utah.edu/units/activities/extraction> [26 July 2004]

Ridley, M. 1999, *Genome: The Autobiography of a Species in 23 Chapters*, Fourth Estate, London

CHAPTER 1

Information on the discovery of Neanderthals, the history of Neanderthal research, and Neanderthal lifestyle and culture was sourced from a number of excellent books on the subject:

Brown, M. H. 1990, *The Search for Eve*, Harper & Row, New York

Constable, G. 1973, *The Emergence of Man: The Neanderthals*, Time Life Books

Lewin, R. 1998, *The Origin of Modern Humans*, Scientific American Library, New York

Shreeve, J. 1995, *The Neandertal Enigma: Solving the Mystery of Modern Human Origins*, Viking, London

Tattersall, I. 1999, *The Last Neanderthal: The Rise, Success, and Mysterious Extinction of Our Closest Human Relatives*, Westview Press, Boulder, Colorado

Trinkaus, E. and Shipman, P. 1993, *The Neandertals: Changing the Image of Mankind*, Alfred A. Knopf, New York

Information about the life of Charles Darwin was taken from the following biography:

White, M. and Gribbon, J. 1995, *Darwin: A Life in Science*, Simon & Schuster, London

Information about Eugene Dubois came from the following book:

Van Oostersee, P. 1999, *Dragon Bones: The Story of Peking Man*, Allen & Unwin, Sydney

Information about Cann, Stoneking and Wilson's research was obtained from the following book:

Shreeve, J. 1995, *The Neandertal Enigma: Solving the Mystery of Modern Human Origins*, Viking, London

Together with their published academic article on the subject:

Cann, R. L., Stoneking, M. and Wilson, A. C. 1987, 'Mitochondrial DNA and human evolution', *Nature*, vol. 325, pp. 31–6

Information about Neanderthal DNA research was obtained from the following published scientific articles:

Krings, M. et al. 1997, 'Neandertal DNA sequences and the origin of modern humans', *Cell*, vol. 90, pp. 19–30

Krings, M. et al. 1999, 'DNA sequence of the mitochondrial hypervariable Region II from the Neandertal type specimen', *PNAS*, vol. 96, pp. 5581–5

Krings, M. et al. 2000, 'A view of Neandertal genetic diversity', *Nature Genetics*, vol. 26, pp. 144–6

Ovchinnikov, I. V. et al. 2000, 'Molecular analysis of Neanderthal DNA from the northern Caucasus', *Nature*, vol. 404, pp. 490–3

Ovchinnikov, I. et al. 2001, 'The isolation and identification of Neanderthal mitochondrial DNA', *Profiles in DNA*, January, pp. 7–9

Together with a number of articles which commented on the Neanderthal DNA research:

Anonymous 1997, 'In our genes?', *The Economist*, 12 July, pp. 77–9

Hawks, J. and Wolpoff, M. 2001, 'Brief communication: Paleoanthropology and the population genetics of ancient genes', *American Journal of Physical Anthropology,* vol. 114, pp. 269–72

Hoss, M. 2000, 'Ancient DNA: Neanderthal population genetics', *Nature,* vol. 404, pp. 453–4

Kahn, P. and Gibbons, A. 1997, 'DNA from an extinct human', *Science,* vol. 277, pp. 176–8

Relethford, J. H. 2001, 'Ancient DNA and the origin of modern humans', *PNAS,* vol. 98, pp. 390–1

Stringer, C. and Ward, R. 1997, 'A molecular handle on the Neanderthals', *Nature,* vol. 388, pp. 225–6

Wolpoff, M. 1998, 'Concocting a divisive theory', *Evolutionary Anthropology,* vol. 7, pp. 1–3

Information on the relationship of Neanderthal DNA to DNA from archaic *Homo sapiens* came from the following article:

Caramelli, D. et al. 2003, 'Evidence for a genetic discontinuity between Neandertals and 24,000 year old anatomically modern humans', *PNAS,* vol. 100, pp. 6593–7

Biographical information about Svante Pääbo came from the following article:

Dickman, S. 1998, 'Svante Pääbo: Pushing ancient DNA to the limit', *Current Biology,* vol. 8, pp. R329–30

Quotes in this chapter came from the following sources:

p. 28 'there isn't a snowball's chance in hell'

p. 28 'if you really want to know where modern humans come from . . .'

from: Shreeve, J. 1995, *The Neandertal Enigma: Solving the Mystery of Modern Human Origins,* Viking, London

p. 32 'a terrific achievement'
from: Stringer, C. and Ward, R. 1997, 'A molecular handle on the Neanderthals', *Nature,* vol. 388, pp. 225–6
p. 32 'an extremely important piece of work'
from: Kahn, P. and Gibbons, A. 1997, 'DNA from an extinct human', *Science,* vol. 277, pp. 176–8
p. 32 'It's not that I want to rain on . . .'
from: Wolpoff, M. 1998, 'Concocting a divisive theory', *Evolutionary Anthropology,* vol. 7, pp. 1–3

CHAPTER 2

Information on the quagga DNA work, and the quagga breeding project, was taken from the following website:

Anonymous 2001, 'The quagga project', *South African Museum website,* <www.museums.org.za/sam/quagga> [21 December 2002]

Together with the following scientific articles:

Higuchi, R. et al. 1984, 'DNA sequences from the quagga, an extinct member of the horse family', *Nature,* vol. 312, pp. 282–4
Higuchi, R. G. et al. 1987, 'Mitochondrial DNA of the extinct quagga: Relatedness and extent of postmortem change', *Journal of Molecular Evolution,* vol. 25, pp. 283–7

And the following commentary:

Jeffreys, A. J. 1984, 'Raising the dead and buried', *Nature,* vol. 312, p. 198

Information about successful ancient DNA extractions from other extinct animals was obtained from the following published articles:

Best, C. H. 1994, 'Genetic analysis of ancient DNA from the hair of the woolly rhinoceros (*Coelodonta antiquitatis*)', *Bridges of the Science Between North America and the Russian Far East*, vol. 45, p. 37

Cooper, A. et al. 1992, 'Independent origins of New Zealand moas and kiwis', *PNAS*, vol. 89, pp. 8741–4

Cooper, A. 2001, 'Complete mitochondrial genome sequences of two extinct moas clarify ratite evolution', *Nature*, vol. 409, pp. 704–7

Christidis, L., Leeton, P. R. and Westerman, M. 1996, 'Were bowerbirds part of the New Zealand fauna?', *PNAS*, vol. 93, pp. 3898–901

Greenwood, A. et al. 1999, 'Nuclear DNA sequences from late Pleistocene megafauna', *Molecular Biology and Evolution*, vol. 16, pp. 1466–73

Greenwood, A. D. et al. 2001, 'A molecular phylogeny of two extinct sloths', *Molecular Phylogenetics and Evolution*, vol. 18, pp. 94–103

Hagelberg, E. et al. 1994, 'DNA from ancient mammoth bones', *Nature*, vol. 370, pp. 333–4

Hanni, C. et al. 1994, 'Tracking the origins of the cave bear *(Ursus spelaeus)* by mitochondrial DNA sequencing', *PNAS*, vol. 91, pp. 12336–40

Hauf, J. et. al. 1995, 'Selective amplification of a mammoth mitochondrial cytochrome B fragment using an elephant specific primer', *Current Genetics*, vol. 27, pp. 486–7

Hoss, M. and Pääbo, S. 1994, 'Mammoth DNA sequences', *Nature*, vol. 370, p. 333

Hoss, M. et al. 1996, 'Molecular phylogeny of the extinct ground sloth *Mylodon darwinii*', *PNAS*, vol. 93, pp. 181–5

Janczewski, D. N. et al. 1992, 'Molecular phylogenetic inference from saber-toothed cat fossils of Rancho La Brea', *PNAS,* vol. 89, pp. 9769–73

Johnson, P. H., Olson, C. B. and Goodman, M. 1985, 'Isolation and characterization of deoxyribonucleic acid from tissue of the woolly mammoth, *Mammuthus primigenius*', *Comparative Biochemistry and Physiology B,* vol. 81, pp. 1045–51

King, S. J., Godfrey, L. R. and Simons, E. L. 2001, 'Adaptive and phylogenetic significance of ontogenetic sequences in *Archaeolemur,* subfossil lemur from Madagascar', *Journal of Human Evolution,* vol. 41, pp. 545–76

Lalueza-Fox, C. et al. 2000, 'Mitochondrial DNA from *Myotragus balearicus,* an extinct bovid from the Balearic Islands', *Journal of Experimental Zoology,* vol. 288, pp. 56–62

Lorielle, O. et al. 2001, 'Ancient DNA analysis reveals divergence of the cave bear, *Ursus spelaeus,* and brown bear, *Ursus arctos,* lineages', *Current Biology,* vol. 11, pp. 200–3

Montagnon, D. et al. 2001, 'Ancient DNA from *Megaladapis edwarsi* (Malagasy subfossil): Preliminary results using partial cytochrome B sequence', *Folia Primatologica,* vol. 72, pp. 30–2

Noro, M. et al. 1998, 'Molecular phylogenetic inference of the woolly mammoth *Mammuthus primigenius,* based on complete sequences of mitochondrial cytochrome B and 12S ribosomal RNA genes', *Journal of Molecular Evolution,* vol. 46, pp. 314–26

Ozawa, T., Hayashi, S. and Mikhelson, V. M. 1997, 'Phylogenetic position of mammoth and Steller's sea cow within Tethytheria demonstrated by mitochondrial DNA sequences', *Journal of Molecular Evolution,* vol. 44, pp. 406–13

Robinson, T. J. et al. 1996, 'Mitochondrial DNA sequences of the extinct blue antelope *Hippotragus leucophaeus*', *Naturwissenschaften,* vol. 83, pp. 178–82

Sorenson, M. D. et al. 1999, 'Relationships of the extinct moa-nalos, flightless Hawaiian waterfowl, based on ancient DNA', *Proceedings of the Royal Society of London Series B Biological Sciences*, vol. 266, pp. 2187–93

Taylor, P. G. 1996, 'Reproducibility of ancient DNA sequences from extinct Pleistocene fauna', *Molecular Biology and Evolution,* vol. 13, pp. 283–5

Trewick, S. A. 1996, 'Flightlessness and phylogeny amongst endemic rails (Aves: Rallidae) of the New Zealand region', *Philosophical Transactions of the Royal Society of London Series B Biological Sciences,* vol. 352, pp. 429–46

Westerman, M. et al. 1999, 'Molecular relationships of the extinct pig-footed bandicoot *Chaeropus ecaudatus* (Marsupialia: Perameloidea) using 12S rRNA sequences', *Journal of Mammalian Evolution,* vol. 6, pp. 271–88

Yang, H., Golenberg, E. M. and Shoshani, J. 1996, 'Phylogenetic resolution within the Elephantidae using fossil DNA sequence from the American mastodon *(Mammut americanum)* as an outgroup', *PNAS,* vol. 93, pp. 1190–4

Background information on the thylacine, thylacine DNA work and the current efforts to clone it was obtained from the following books, website and television documentary:

Anonymous 2002, 'Australia's thylacine', *Australian Museum Online,* <www.austmus.gov.au/thylacine> [4 October 2002]

'End of extinction: Cloning the Tasmanian tiger', Discovery Channel [19 Jan 2004]

Owen, D. 2003, *Thylacine: The Tragic Tale of the Tasmanian Tiger,* Allen & Unwin, Sydney

Paddle, R. 2001, *The Last Tasmanian Tiger: The History and*

Extinction of the Thylacine, Cambridge University Press, Cambridge

Together with the following scientific articles:

Krajewski, C. et al. 1992, 'Phylogenetic relationships of the thylacine (Mammalia, Thylacinidae) using dasyuroid marsupials—evidence from cytochrome B DNA sequences', *Proceedings of the Royal Society of London Series B Biological Sciences*, vol. 250, pp. 19–27

Krajewski, C., Buckley, L. and Westerman, M. 1997, 'DNA phylogeny of the marsupial wolf resolved', *Proceedings of the Royal Society of London Series B Biological Sciences*, vol. 264, pp. 911–17

Thomas, R. H. et al. 1989, 'DNA phylogeny of the extinct marsupial wolf', *Nature*, vol. 340, pp. 465–7

General information on genome projects, including the human genome project, came from the following book:

Sulston, J. and Ferry, G. 2002, *The Common Thread: A Story of Science, Politics, Ethics and the Human Genome*, Bantam Press, London

Background information on Ice Age animals, and information about the work to clone a mammoth, was taken from the following books and websites:

Anonymous 2001, 'Will mammoths walk again?', *Discovery Channel website*, <www.exn.ca/mammoth/Cloning.cfm> [5 January 2004]

Anonymous 2003, 'Live cells' found in frozen mammoth', *The Geological Society (UK) website*, <www.geolsoc.org.uk/template.cfm?name=mammoth2> [5 January 2004]

Cohen, C. 2002, *The Fate of the Mammoth: Fossils, Myth and History*, University of Chicago Press, Chicago

Gasperini, B. 2003, 'Mammoth clone: Science, or simply fiction?', *Discovery Channel website*, <dsc.discovery.com/convergence/lanofmammoth/dispatches/clonezone.html> [5 January 2004]

Hehner, B. and Hallett, M. (illustrator) 2001, *Ice Age Mammoth: Will This Ancient Giant Come Back to Life?*, Scholastic

Martin, P. S. and Wright, H. E. Jr. 1967, *Pleistocene Extinctions: The Search for a Cause*, Yale University Press, New Haven

Stone, R. 2001, *Mammoth: The Resurrection of an Ice Age Giant*, Perseus Publishing, Cambridge, Massachusetts

Sutcliffe, A. J. 1985, *On the Track of Ice Age Mammals*, Harvard University Press, Cambridge, Massachusetts

Ward, P. D. 1997, *The Call of Distant Mammoths: Why the Ice Age Mammals Disappeared*, Copernicus, New York

The following book was also used in preparing this chapter:

Dawkins, R. 2000, *The Blind Watchmaker*, Penguin Books, London

CHAPTER 3

The background information about dinosaurs and the dinosaur era in this chapter was sourced from the following books:

Cadbury, D. 2000, *The Dinosaur Hunters: A Story of Scientific Rivalry and the Discovery of the Prehistoric World*, Fourth Estate, London

Haines, T. 1999, *Walking with Dinosaurs: A Natural History*, BBC Worldwide Limited, London

Information about the formulation of plans to extract dinosaur DNA, and of Schweitzer's initial work, came from the following book and article:

DeSalle, R. and Linley, D. 1997, *The Science of Jurassic Park and the Lost World: Or, How to Build a Dinosaur*, Basic Books, New York
Morell, V. 1993, 'Dino DNA: The hunt and the hype', *Science*, vol. 261, pp. 160–2

Information about plant compression fossil DNA research came from the following scientific articles:

Golenberg, E. M. et al. 1990, 'Chloroplast DNA sequence from a Miocene *Magnolia* species', *Nature*, vol. 344, pp. 656–8
Golenberg, E. M. 1991, 'Amplification and analysis of Miocene plant fossil DNA', *Philosophical Transactions of the Royal Society of London Series B Biological Sciences*, vol. 333, pp. 419–27
Golenberg, E. M. 1994, 'DNA from plant compression fossils' in *Ancient DNA*, eds B. Herrmann and S. Hummel, Springer Verlag, New York, pp. 237–56
Manen, J. F. et al. 1995, 'Chloroplast DNA sequences from a Miocene diatomite deposit in Ardeche (France)', *C R Acad Sci Paris, Life Sciences*, vol. 318, pp. 971–5
Soltis, P. S. et al. 1992, 'An *rbcL* sequence from a Miocene *Taxodium* (bald cypress)', *PNAS*, vol. 89, pp. 449–51

Together with this commentary article:

Pääbo, S. and Wilson, A. C. 1991, 'Miocene DNA sequences—a dream come true?', *Current Biology*, vol. 1, pp. 45–6

Background information on amber came from the following article:

Langenheim, J. H. 1990, 'Plant resins', *American Sci*, vol. 78, pp. 16–24

Together with the following book, which also comments on the DNA work from insects in amber, and on dinosaur DNA work:

DeSalle, R. and Linley, D, 1997, *The Science of Jurassic Park and the Lost World. Or, How to Build a Dinosaur*, Basic Books, New York

Further information about the DNA work on insects in amber came from the following articles:

Cano, R. J. et al. 1992, 'Isolation and partial characterisation of DNA from the bee *Proplebeia dominicana* (Apidae: Hymenoptera) in 25–40 million year old amber', *Med Sci Research*, vol. 20, pp. 249–51

Cano, R. J. et al. 1992, 'Enzymatic amplification and nucleotide sequencing of portions of the 18S rRNA gene of the bee *Proplebeia dominicana* (Apidae: Hymenoptera) isolated from 25–40 million year old Dominican amber', *Med Sci Research*, vol. 20, pp. 619–22

DeSalle, R. et al. 1992, 'DNA sequences from a fossil termite in Oligo–Miocene amber and their phylogenetic implications', *Science*, vol. 257, pp. 1933–6

Information about dinosaur DNA research was obtained from the following scientific articles:

An, C. C. et al. 1995, 'Molecular cloning and sequencing of the 18S rDNA from specialized dinosaur egg fossil found in Xixia Henan, China', *Acta Scientiarum Naturalium Universitatis Pekinensis*, vol. 31, pp. 140–7

Li, Y. et al. 1995, 'DNA isolation and sequence analysis of dinosaur DNA from Cretaceous dinosaur egg in Xixia Henan, China', *Acta Scientiarum Naturalium Universitatis Pekinensis*, vol. 31, pp. 148–52

Woodward, S. R. et al. 1994, 'DNA sequence from Cretaceous period bone fragments', *Science*, vol. 266, pp. 1229–32

Zhang, Y. and Fang, X. 1995, 'A late Cretaceous dinosaur egg with preserved genetic information from Xixia Basin, Henan, China: Structure, mineral-chemical and taphonomical analyses', *Acta Scientiarum Naturalium Universitatis Pekinensis*, vol. 31, pp. 129–39

Zou, Y. P. et al. 1995, 'Ancient DNA in late Cretaceous dinosaur egg from Xixia County, Henan Province', *Chinese Science Bulletin*, vol. 40, pp. 856–60

Together with the following commentary:

Gibbons, A. 1994, 'Possible dino DNA find is greeted with skepticism', *Science*, vol. 266, p. 1159

Information about the doubts raised about the super-ancient DNA work was obtained from the following published articles:

Hedges, S. B. and Schweitzer, M. H. 1995, 'Detecting dinosaur DNA', *Science*, vol. 268, p. 1191

Sidow, A. et al. 1991, 'Bacterial DNA in Clarkia fossils', *Philosophical Transactions of the Royal Society of London Series B Biological Sciences*, vol. 333, pp. 429–33

Wang, H. 1996, 'Re-analysis of DNA sequence data from a dinosaur egg fossil unearthed in Xixia of Henan Province', *Yi Chuan Xue Bao*, vol. 23, pp. 183–9

Wang, H. L. et al. 1997, 'Re-analysis of published DNA sequence amplified from Cretaceous dinosaur egg fossil', *Molecular Biology and Evolution*, vol. 14, pp. 589–91

Woodward, S. R. 1995, 'Detecting dinosaur DNA', *Science*, vol. 268, p. 1194

Yin, Z. et al. 1996, 'Sequence analysis of the cytochrome B gene fragment in a dinosaur egg', *Yi Chuan Xue Bao,* vol. 23, pp. 190–5

Young, D. L. et al. 1995, 'Testing the validity of the cytochrome B sequence from Cretaceous period bone fragments as dinosaur DNA', *Cladistics,* vol. 11, pp. 199–209

The following lectures and radio interview were also used as significant sources of information for this chapter:

Alan Cooper, Joint presentation between the Royal Society of New Zealand and Victoria University, 15 April 2003 [Public Lecture]

Kim Hill with Professor Alan Cooper, National Radio, 12 April 2003, Radio New Zealand Ltd, Wellington [Radio Broadcast]

David Lambert, Lecture at Massey University, 2003 [Public Lecture]

Information about the possibility of 'breeding' a dinosaur came from the following article:

Ridley, M. 2000, 'Will we clone a dinosaur?', *Time,* vol. 155, pp. 94–5

Quotes in this chapter came from the following sources:

p. 81 'like nothing we've ever seen before . . .'
from: Gibbons, A. 1994, 'Possible dino DNA find is greeted with skepticism', *Science,* vol. 266, p. 1159

p. 86 'The way to get around this . . .'

p. 87 'But there's a slight problem . . .'

p. 87 'by no means does that mean . . .'

p. 88 'We ourselves are incredibly dirty . . .'

p. 88 'in sweat on your skin . . .'

p. 89 'If you put DNA in water . . .'
p. 89 'In permafrost conditions . . .'
from: Alan Cooper, Joint Presentation between the Royal Society of New Zealand and Victoria University, 15 April 2003, [Public Lecture]
p. 87 'you've got enough there . . .'
from: Kim Hill with Professor Alan Cooper, National Radio, 12 April 2003, Radio New Zealand Ltd, Wellington [Radio Broadcast]

CHAPTER 4

The following extremely interesting and informative interview and lecture were used as significant sources of information throughout this chapter:

Kim Hill with Professor Alan Cooper, National Radio, 12 April 2003, Radio New Zealand Ltd, Wellington [Radio Broadcast]
David Lambert, Lecture at Massey University, 2003 [Public Lecture]

Information on the history of moa discoveries and research came from the following books:

Wolfe, R. 2003, *Moa: The Dramatic Story of the Discovery of a Giant Bird*, Penguin Books, Auckland
Worthy, T. H. and Holdaway, R. N. 2002, *The Lost World of the Moa: Prehistoric Life of New Zealand*, Indiana University Press, Bloomington, Indiana

Information on the meaning of species, genus and family came from the following websites:

Anonymous 1998, 'What is a genus and what is a species?', *University of Florida website*, <http://aquat1.ifas.ufl.edu/genspe.html> [20 July 2004]

Anonymous 2004, 'family', *Dictionary.com*, <http://dictionary.reference.com/search?q=family&r=67> [20 July 2004]

Information on the research into the origin and relationship of the moa and kiwi came from the following articles:

Cooper, A. et al. 1992, 'Independent origins of New Zealand moas and kiwis', *PNAS*, vol. 89, pp. 8741–4

Cooper, A. et al. 1993, 'Evolution of the moa and their effect on the New Zealand flora', *TREE*, vol. 8, pp. 433–7

Cooper, A. et al. 2001, 'Complete mitochondrial genome sequences of two extinct moas clarify ratite evolution', *Nature*, vol. 409, pp. 704–7

Information on the research into the number of species of moas came from the following articles:

Bunce M. et al. 2003, 'Extreme reversed sexual size dimorphism in the extinct New Zealand moa *Dinornis*', *Nature*, vol. 425, pp. 172–5

Huynen, L., Millar, C. D. and Lambert, D. M. 2002, 'A DNA test to sex ratite birds', *Molecular Ecology*, vol. 11, pp. 851–6

Huynen, L. et al. 2003, 'Nuclear DNA sequences detect species limits in ancient moa', *Nature*, vol. 425, pp. 175–8

Together with email correspondence with David Lambert [23 September 2004]

Quotes in this chapter came from the following sources:

p. 102 'absolute rabid caver'

p. 102 'And of course . . .'

p. 102 'They're actually quite good crowbars . . .'

p. 102 'Reading this, I suddenly worked out . . .'

p. 102 'They've been an evolutionary puzzle . . .'
p. 104 'We do know that ratites can swim . . .'
p. 106 'Technically, if you look at the relationships . . .'
from: Kim Hill with Professor Alan Cooper, National Radio, 12 April 2003, Radio New Zealand Ltd, Wellington [Radio Broadcast]
p. 111 'we understand a lot . . .'
p. 111 'one of the things we're working on now . . .'
p. 111 'we think now that we can amplify the genes . . .'
from: David Lambert, Lecture at Massey University, 2003 [Public Lecture]

CHAPTER 5

Information on the definition of epidemic and pandemic came from the following websites:

Anonymous, no date, 'pandemic' *Webster's Online Dictionary,* <http://www.webster-dictionary.org/definition/pandemic> [21 July 2004]

Anonymous, no date, 'epidemic' *Webster's Online Dictionary,* http://www.webster-dictionary.org/definition/epidemic [21 July 2004]

Information on the history of the Black Death came from the following books:

Naphy, W. and Spicer, A. 2001, *The Black Death: A History of Plagues 1345–1730,* Tempus Publishing, Stroud, Gloucestershire

Ziegler, P. 1997, *The Black Death,* Sutton Publishing, Gloucestershire

Together with the following article, which also provided information about Scott and Duncan's work, and Raoult's work:

Anonymous 2001, 'Ring a ring o'roses', *New Scientist*, 21 November, pp. 33–7

Information about Raoult's research was also obtained from the following articles:

Drancourt, M. G. et al. 1998, 'Detection of 400-year-old *Yersinia pestis* DNA in human dental pulp: An approach to the diagnosis of ancient septicaemia', *PNAS*, vol. 95, pp. 12637–40

Drancourt, M. and Raoult, D. 2002, 'Molecular insights into the history of plague', *Microbes and Infection*, vol. 4, pp. 105–9

Raoult, D. et al. 2000, 'Molecular identification by "suicide PCR" of *Yersinia pestis* as the agent of medieval Black Death', *PNAS*, vol. 97, pp. 12800–3

And the following commentary article:

Wasson, K. and O'Neill, M. D. 2003, 'Suicide PCR identifies *Yersinia pestis* DNA in Black Death victims', *Applied BioSystems BioBeat Newsletter*, <www.appliedbiosystems.com/biobeat/suicide/> [4 December 2003]

The information on Alan Cooper's work on plague DNA came from the following website:

MacKenzie, D. 2003, 'Case reopens on Black Death cause', *Discovery Channel UK website*, <http://www.discoverychannel.co.uk/newscientist/week02/article03.shtml> [10 December 2003]

Most of the background information in this chapter regarding Columbus's life and journeys was taken from an excellent biography of Columbus:

Phillips, W. D. Jr. and Phillips, C. R. 1992, *The Worlds of Christopher Columbus*, Cambridge University Press, Cambridge

The information on the tuberculosis DNA work on pre-Columbian mummies came from the following articles:

Arriaza, B. T. et al. 1995, 'Pre-Columbian tuberculosis in northern Chile: Molecular and skeletal evidence', *American Journal of Physical Anthropology*, vol. 98, pp. 37–45

Braun, M. J., Cook, D. and Pfeiffer, S. 1998, 'DNA from *Mycobacterium tuberculosis* complex identified in north American, pre-Columbian human skeletal remains', *Journal of Archaeological Science*, vol. 25, pp. 271–7

Salo, W. L. et al. 1994, 'Identification of *Mycobacterium tuberculosis* DNA in a pre-Columbian Peruvian mummy', *PNAS*, vol. 91, pp. 2091–4

Together with the following commentary articles:

Charatan, F. B. 1994, 'Peruvian mummy shows that TB preceded Columbus', *BMJ*, vol. 308, p. 808

Morell, V. 1994, 'Mummy settles TB antiquity debate', *Science*, vol. 263, pp. 1686–7

The majority of the historical information on the 1918 flu epidemic and on Taubenberger's work comes from the following excellent and entertaining book:

Kolata, G. 1999, *Flu: The Story of the Great Influenza Pandemic of 1918 and the Search for the Virus that Caused It*, Pan Books, London

Additional information on Taubenberger's research came from the following articles:

Basler, C. F. et al. 2001, 'Sequence of the 1918 pandemic influenza virus nonstructural gene (NS) segment and characterization of the recombinant viruses bearing the 1918 NS genes', *PNAS*, vol. 98, pp. 2746–51

Reid, A. H. et al. 1999, 'Origin and evolution of the 1918 'Spanish' influenza virus hemagglutinin gene', *PNAS*, vol. 96, pp. 1651–6

Reid, A. H. et al. 2000, 'Characterization of the 1918 "Spanish" influenza virus neuraminidase gene', *PNAS*, vol. 97, pp. 6785–90

Taubenberger, J. K. et al. 1997, 'Initial genetic characterization of the 1918 "Spanish" influenza virus', *Science*, vol. 275, pp. 1793–6

Information on the ANU research came from the following articles:

Gibbs, M. J., Armstrong, J. S. and Gibbs, A. J. 2001, 'Recombination in the hemagglutinin gene of the 1918 "Spanish Flu"', *Science*, vol. 293, pp. 1842–5

Gibbs, M. J., Armstrong, J. S. and Gibbs, A. J. 2001, 'The hemagglutinin gene, but not the neuraminidase gene, of "Spanish flu" was a recombinant', *Philosophical Transactions of the Royal Society of London Series B Biological Sciences*, vol. 356, pp. 1845–55

Together with the following commentary article:

Jackson, C. 2001, 'ANU stuns world with killer-flu theory', *The Canberra Times*, September 8, pp. 1–2

Quotes in this chapter came from the following sources:

p. 124 'We believe that we can end the controversy . . .'
from: Anonymous, 2001, 'Ring a ring o'roses' *New Scientist*, 24 November, pp. 35–7

p. 125 'We cannot rule out *Yersinia pestis* as the cause . . .'
from: MacKenzie, D. 2003, 'Case reopens on Black Death

cause', *Discovery Channel UK website*, <http://www.discoverychannel.co.uk/newscientist/week02/article03.shtml> [10 December 2003]

p. 132 'This provides the most specific evidence . . .'
from: Salo, W. L. et al. 1994, 'Identification of *Mycobacterium tuberculosis* DNA in a pre-Columbian Peruvian mummy', *PNAS*, vol. 91, pp. 2091–4

p. 138 'This is a detective story . . .'

p. 142 'armies of white blood cells and fluids . . .'
from: Kolata, G. 1999, *Flu: The Story of the Great Influenza Pandemic of 1918 and the Search for the Virus that Caused It*, Pan Books, London

CHAPTER 6

The bulk of the information in this chapter about the Russian revolution, and the lives of Anastasia and Anna Anderson, was taken from two excellent biographies:

Klier, J. and Mingay, H. 1995, *The Quest for Anastasia: Solving the Mystery of the Lost Romanovs*, Smith Gryphon Publishers, London

Kuth, P. 1985, *Anastasia: The Life of Anna Anderson*, Fontana Paperbacks, Great Britain

Information on the DNA work on the Russian royal family and on Anna Anderson also came from the following book and articles:

Gill, P. et al. 1994, 'Identification of the remains of the Romanov family by DNA analysis', *Nature Genetics*, vol. 6, pp. 130–5

Klier, J. and Mingay, H. 1995, *The Quest for Anastasia: Solving the Mystery of the Lost Romanovs*, Smith Gryphon Publishers, London.

Ivanov, P. L. et al. 1996, 'Mitochondrial DNA sequence heteroplasmy in the Grand Duke of Russia Georgij Romanov establishes the authenticity of the remains of Tsar Nicholas II', *Nature Genetics*, vol. 12, pp. 417–21

Stoneking, M. et al. 1995, 'Establishing the identity of Anna Anderson Manahan', *Nature Genetics*, vol. 9, pp. 9–10

Together with the following commentary articles:

Anonymous 1994, 'Anastasia and the tools of justice', *Nature Genetics*, vol. 8, pp. 205–6

Anonymous 1996, 'Romanovs find closure in DNA', *Nature Genetics*, vol. 12, p. 339

Schweitzer, R. 1995, 'Anastasia and Anna Anderson', *Nature Genetics*, vol. 9, p. 345

Information on standard paternity testing methods came in part from a personal email from Jean-Jacques Cassiman [23 July 2004]

Information about a recent Anastasia claimant came from the following article:

Anonymous 2002, 'Century-old Georgian woman claims Russian tsar's fortune', *ABC Online*, <http://abc.net.au/cgi-bin/common/printfriendly.pl?http://abc.net.au/news/newsitems/s579335.htm> [12 June 2002]

Information on the DNA work on the *Titanic*'s 'unknown child' came from the following sources:

Anonymous 2002, 'Titanic's "unknown child" identification continues. *Lakehead University website* <http://www.lakehead.ca/~eventswww/titanic_release.html> [9 December 2004]

Anonymous, 2004, 'Scientists identify Titanic's "unknown child"', *CTV.ca*, <http://www.ctv.ca/servlet/ArticleNews/story/CTVNews/1036603301117_32012501?s_name=&no_ads> [9 December 2004]

Carter, L. 2002, 'Titanic's baby victim identified', *BBC News website* <http://news.bbc.co.uk/1/hi/world/americas/2413895.stm> [9 December 2004]

Laydier, K. 2002, 'Lakehead lab probes Titanic DNA mystery', *The Chronicle Journal* <http://www.ancientdna.com/jan27-02a.htm> [9 December 2004]

Legge, L. 2002, 'Experts narrow Titanic mystery', *Halifax Chronicle Herald*, <www.canoe.ca/CNEWSFeatures0202/27_titanic-par.html> [9 December 2004]

Parr, R. et al. 'Working towards genetic analysis of an "unknown child" from the 1912 RMS Titanic disaster', *Poster presented at the 6th international conference on Ancient DNA and Associated Biomolecules held in Tel Aviv, Israel, 21–25 July 2002*, <http://www.ancientdna.com/Titanic.htm> [9 December 2004]

'Titanic's "unknown child" identified', Lakehead University, 6 November 2002 [Press Release]

Titley K. et al. 2004, 'The *Titanic* disaster: dentistry's role in the identification of an "Unknown Child"', *J Can Dent Assoc*, vol. 70, pp. 24–8

Quotes in this chapter came from the following sources:

p. 155 'She in no way resembles the true Grand Duchess Anastasia . . .'

p. 158 'That's enough of this Romanov business . . .'

p. 164 'proves virtually beyond doubt that five of the nine skeletons . . .'

p. 165 'All I want is the truth . . .'
from: Klier, J. and Mingay, H. 1995, *The Quest for Anastasia: Solving the Mystery of the Lost Romanovs*, Smith Gryphon Publishers, London

p. 172 'The unknown child is now a known child . . .'
from: 'Titanic's "unknown child" identified', Lakehead University, 6 November 2002 [Press Release]

CHAPTER 7

The bulk of the background information in this chapter about the lives of Louis XVII and Karl Naundorff was taken from two excellent biographies:

Cadbury, D. 2002, *The Lost King of France: Revolution, Revenge and the Search for Louis XVII*, Fourth Estate, London

Madol, H. R. 1930, *The Shadow King: The Life of Louis XVII of France and the Fortunes of the Naundorff–Bourbon Family*, George Allen & Unwin Ltd, London

Extra background information about the events of the French Revolution was taken from the following book:

Jones, C. 1994, *The Cambridge Illustrated History of France*, Cambridge University Press, Cambridge

Information about the DNA testing of the remains of Naundorff and Louis XVII was sourced primarily from two articles:

Jehaes, E. et al. 1998, 'Mitochondrial DNA analysis on remains of a putative son of Louis XVI, King of France, and Marie-Antoinette', *European Journal of Human Genetics*, vol. 6, pp. 383–95

Jehaes, E. et al. 2001, 'Mitochondrial DNA analysis of the putative heart of Louis XVII, son of Louis XVI and

Marie-Antoinette', *European Journal of Human Genetics*, vol. 9, pp. 185–90

Together with phone [24 October 2002] and email [15 January 2004, 23 July 2004] conversations with Jean-Jacques Cassiman

Some information about the DNA testing is also taken from:

Cadbury, D. 2002, *The Lost King of France: Revolution, Revenge and the Search for Louis XVII*, Fourth Estate, London

Quotes in this chapter came from the following sources:

p. 173 'How ugly everything is here'
p. 181 'exchanged recollections which alone . . .'
p. 183 'My dearly beloved sister . . .'
p. 183 'assure you of the existence of your illustrious brother . . .'
p. 183 'As a servant of your illustrious father . . .'
from: Madol, H. R. 1930, *The Shadow King: The Life of Louis XVII of France and the Fortunes of the Naundorff–Bourbon Family*, George Allen & Unwin Ltd, London

All quotes from Jean-Jacques Cassiman are from personal phone [24 October 2002] and email [15 January 2004, 23 July 2004] conversations

CONCLUSION

Information on the future of ancient DNA research, and all quotes from Alan Cooper, were taken from the following radio interview and lecture:

Alan Cooper, Joint presentation between the Royal Society of New Zealand and Victoria University, 15 April 2003 [Public Lecture]

Kim Hill with Professor Alan Cooper, National Radio, 12 April 2003, Radio New Zealand Ltd, Wellington [Radio Broadcast]

Information on the future of ancient DNA research was also taken from an email conversation [11 February 2004] with Alan Cooper

The following article was also used as a source of information for the conclusion:

Zink, A. R. et al. 2002, 'Molecular analysis of ancient microbial infections', *FEMS Microbiology Letters*, vol. 213, pp. 141–7

INDEX

adenine (A), 2, 30, 140
Africa
  common human ancestor, 27
  'Out of Africa' hypothesis, 23, 28
amber
  ancient DNA, 76–8
  Cretaceous period, 77
  definition, 77
  extraction process, 78–80
American mastodons
  DNA from, 41
Anastasia
  Anna Anderson, 145, 152–5
  fate of, 144, 169
  Nicholas II, daughter of, 143
  royal grave, search for, 158–60
  rumoured murder of, 150–2
Ancient Biomolecules Centre, 88
ancient DNA
  dinosaurs *see* dinosaur DNA
  extinct species, from, 35
  future research, 194–8
  permafrost core samples, from, 197
  plant evidence, 74–6
  replication of extractions, 85–6
  research, 2, 6, 194–8
  temporal limits, 89–90
Anna Anderson
  Anastasia, claiming to be, 145, 152–5
  DNA identification, 165–7
  identity of, 167–9
  legal recognition, quest for, 156–8
  supporters of, 155–6
Anning, Mary, 67
anti-contamination measures, 88
Archaeopteryx, 71
Archer, Mike, 41, 45–6, 49
Ardèche fossils, 76
Armstrong, John, 141
Aufderheide, Arthur, 131–3
Australia, arrival of humans in, 42
Avodinin, Alexander, 158–9

Bastille, storming of, 174
Berezovka mammoth, 56
birds
  descendants of dinosaurs, 84, 89
  sex chromosomes, 108
Black Death, 112, 114, 115–16
  causes of, 114
  DNA and dental pulp, 122–5
  duration of pandemic, 119
  spread of, 118–19

Black Death *continued*
  symptoms, 117–18
  the plague and, 120–22, 124, 195
blue antelopes, DNA from, 41
Botkin, Dr Eugene Sergeyevitch, 150, 155, 159, 165
Botkin, Gleb, 155–7, 165
Botkin, Tatiana, 155
Braun, Mark, 132
breeding project
  quagga, 64–5
Buigues, Bernard, 61
Burkhart, Susan, 166

Cambridge University Zoological Museum, 103
Cann, Rebecca, 26–8
Canterbury Museum, 97
Cassiman, Professor Jean-Jacques, 186–7, 189, 190–193
cave bears
  DNA from, 41
  extinction, 52
  Ice Age megafauna, 51
Chagas disease, 195
chromosomes, 2–4
Clarkia fossil beds, 74–6, 85
cloning
  dinosaurs, 82–3
  Dolly the sheep, 60
  ethics and, 49–50, 63–4
  future ancient DNA research, 196
  mammoths, 57–62
  Tasmanian tiger, 45–9
Colenso, William, 94, 96
Columbus
  tuberculosis and, 112, 114, 126–33
Comte d'Artois, 177
Comte de Provence, 177
Cook, Della, 132
Cooper, Alan, 1–2, 7, 86–7, 89, 102–6, 111, 124–5, 194, 196, 197
Cracraft, Joel, 106–7
Cretaceous period, 68, 69
  amber, 77
cytosine (C), 2, 30, 140

d'Angouleme, Duchess, 102–3, 189
Darwin, Charles, 93
  The Origin of the Species, 16
  theory of natural selection, 12, 13, 16
de Bourbon, Don Carlos, 190
De Quelin, Monsieur, 189
Delorme, Philippe, 189
DeSalle, Rob, 78–9
Dima, 56–7
Dinornis, 95
Dinornis giganteus, 99
dinosaur DNA
  extraction, 80–2
  human DNA, similarity to, 84
  new frontier, 72–3
  sceptical scientists, 83–6
  sources, 73–4
dinosaurs
  cloning, 82–3
  extinction, 38, 52, 68, 70–1, 89
  family tree, 69–70
  fossil discoveries, 13, 67–8
  genome, 82
  naming of, 68, 93
  public fascination with, 66
  timeline, 68–9
diphtheria, 195
diprotodon, 42
DNA
  adenine (A), 2, 30, 140
  ancient *see* ancient DNA
  bases, 2, 30
  basics, 2–5
  cytosine (C), 2, 30, 140
  decay of, 89
  dirt, in, 197
  double helix, 3, 30
  evolution of, 27, 197

guanine (G), 2, 30, 140
human *see* human DNA
inheritance of, 4, 27
limited life of, 88–90
mutations, 27
sequencers, 31
strands, 30
stray, 78–9, 84, 86, 87
thymine (T), 2, 30, 140
water, deterioration in, 85, 89
DNA extraction
anti-contamination measures, 88
contamination, 87
extinct species, from, 39–46
humans, from, 26–8
Neanderthals, from, 29–31, 33
plant fossils, from, 74–6
process of, 5–6, 30–1
quagga, from, 39–41
DNA fingerprinting, 143
Dolly the sheep, 60
double helix, 3, 30
Downs, James, 139
Dubois, Eugene, 17
duc d'Orleans, 177
Duc de Bauffremont, 190
Dumont, Eduard, 190
Duncan, Christopher, 121–2, 124

E. coli, 195
epidemics, 114
indigenous Americans, in, 130
ethics
cloning and, 49–50, 63–4
Etosha National Park, 64
evolution
human *see* human evolution
process of, 12
theories of, 12, 13
extinct species
ancient DNA research, future of, 196
bringing back to life, 35, 36–9
dinosaurs, 38, 52, 68, 70–1, 89
extracting DNA from, 39–46
Neanderthals, 24, 34
Tasmanian tiger, 37, 41
traditional analyses, [illegible]
woolly mammoth, 37
extinction, causes of, 52–4

family, 98
fossils
dinosaurs, 13, 67–8
discoveries, 13
human ancestors and, 23–5
leaf compression fossils, 74–6
New Zealand, in, 100
French Revolution, 173–7
Fuhlrott, Carl, 11, 13–14

GenBank, 81
gender, human bones, 108
genes, 4
genome
dinosaurs, 82
thylacine, 47
genome project, 47
genus, 98
giant deer, 51
Gibbs, Adrian, 141
Gibbs, Mark, 141
Gill, Peter, 160, 162, 165–6
global warming, 51, 196
Golenberg, Edward, 74
gomphotheres, 55
Gondwana, 69, 100–1, 104
Goto, Kazufumi, 58–62
Grand Duke Georgij, 164
guanine (G), 2, 30, 140
haemagglutinin, 141

Harris, John W., 91
Homo erectus
fossils, 25
human ancestor, 17–18, 23–5

Homo erectus continued
'racial' differences, 18
Homo sapiens, 18
Hultin, Johan, 140
human DNA
Neanderthal DNA and, 31, 33
similarity of, 27
human evolution
Christian tradition, 11–12
common ancestor, 27
debate about, 22–5
fossilised ancestors, searches for, 16
Neanderthals, place in, 10, 22–5, 32–5
similarity of DNA in humans, 27
Human Genome Project, 47
humans
Australia, arrival in, 42
DNA extraction, 26–8
soul, 12
hybrid animals, infertility, 59

Ice Age, 42, 50–2
animals, research on, 196
extinction of megafauna, 53–4
mammoths, 50–2
megafauna, 50–2, 196
infectious diseases, 113
future ancient DNA research, 195
history of, 195
influenza pandemic (1918), 112, 114, 134–8, 195
DNA and, 138–42
origin and cause, 135, 137, 138
severity of, 114, 134
symptoms, 135
vaccinations, 136
influenza, research into, 137
insects
ancient DNA, 76–8
extraction process, 78–80
iridium, 71
Iritani, Akira, 60

Jarkov the mammoth, 61
Java, 17
Jurassic period, 68, 69
amber, 77

Khatanga, 61
King Louis XVI of France, 173–7
execution, 177
Temple of Paris, in, 176
trial, 176
King Louis XVII *see* Louis-Charles
King Louis XVIII, 180, 189
kiwis, 95, 99, 100, 104
origin, 106
Klier, John, 168
Kobayashi, Kazutoshi, 60
Kolata, Gina, 138, 142
Krings, Matthias, 30–1

Lambert, David, 107–11
Laurasia, 69
Lazarev, Pyotr, 60
leaf compression fossils, 74–6, 85
lemurs, DNA from, 41
Lenin, 149–50
leprosy, ancient DNA extraction, 195
Logachev, Anatoly, 56
Louis-Charles, 173–8, 193
birth of, 173
burial, 179
death of, 178, 191
DNA extraction from heart of, 191
DNA of relatives, 187–8
heart, removal of, 179
illness, 178
imprisonment, 177–8
locating heart of, 189–90
mystery about, 179–80
post-mortem, 179

proclaimed regent, 177
tuberculosis, 178

malaria, 195
mammals
dinosaur extinction, following, 71
groups, 42
mammoths *see also* woolly mammoths
Berezovka mammoth, 56
causes of extinction, 53–4
cloning, 57–62
Dima, 56–7
discovery of remains of, 55–6
DNA from, 41
elephants, relation to, 55
extinction, 52
habitat, 63
Ice Age and, 50–2
mammoth/elephant hybrid, 59
'mammoth steppe', 63
permafrost, preservation in, 54
temperament, 63
Manahan, Jack, 157–8
Maori middens, 95
Mappin, Frank Crossley, 99
Mappin's moa, 99
Marie-Antoinette, 173, 187, 192
execution, 178
trial, 178
marsupials, 42
mastodons
DNA from, 41
mammoths and, 55
Maucher, Carl, 168
megafauna, 42
causes of extinction, 53–4
extinction, 42–3, 51, 196
Ice Age, in, 50–2
Mesozoic era, 68
meteorites
extinction of dinosaurs, 70–1
iridium, 71
Mikhelson, Viktor, 57
moa-nalos, DNA from, 41
moas, 6, 37
ancient DNA research, 41
bones, 91, 94–5, 97, 100
colour, 111
DNA, 102–3, 197
evolutionary history, 100–6
extinction, 96
families, 98
future of research about, 111–12
genera, 98
genetic comparisons, 104
habitats, 99
hunting, 93–5
largest, 98–9
origins, 95–100
questions about, 90
ranges, 99
sexual dimorphism, 107–11
size, 98–9
species of, 96, 98, 106–10
Mol, Dirk Jan, 61
monotremes, 42
Montpellier, 122
'Multiregional Evolution', 25, 33
Museum of New Zealand Te Papa Tongarewa, 103

National Museum of New Zealand, 103
natural history, 13
Natural History Museum London, 23, 93
natural selection, theory of, 12, 13, 16
Naundorff, Karl Wilhelm
death of, 185
DNA analysis, 186–8
Louis-Charles, claiming to be, 180–5, 191–3
Neander Valley, 11–13, 16, 30
Neanderthals
ancient DNA, 6
brain, 21

Neanderthals *continued*
culture, 18–21
dead, burial of, 21
depiction of, 10
discovery of, 11–14
DNA extraction, 29–31, 33
extinction, 10, 24, 34
eyebrow ridges, 9–10, 19
fate of, 22
first skeleton, discovery of, 10, 11–14
great debate about, 14–16, 22–5
human ancestors, whether, 10, 22–5, 32–5
human evolution, place in, 10
lifestyle, 18–21
modern-day humans, relationship to, 22
myths about, 19
naming of, 16
origin of, 18
questions about, 16
skull, 19, 20
tools, 21
New Zealand
flora and fauna, 101
fossils, 100
New Zealand rail species, DNA from, 41

Origin of the Species, The, 16
Otago Museum, 103
'Out of Africa' hypothesis, 23, 28
Ovchinnikov, Igor, 33
Owen, Richard, 68, 91–2, 94–6
Ozawa, Tomowo, 57

Pääbo, Svante, 29–31, 85
Pachyornis mappini, 99
Paglicci cave, 34
Pålsson, Gösta Leonard, 170
pandemics, 114
Pangaea, 68, 69, 100
Panula, Eino Viljami, 172
Parr, Ryan, 170
paternity, determining, 161
PCR *see* polymerase chain reaction
Pelletan, Dr Philippe-Jean, 179, 189
permafrost
ancient DNA preserved in, 54, 89, 196
core samples, 197
definition, 54
preservation and, 54, 89
Siberian, 197
Petrie, Hans, 186
Pfeiffer, Susan, 132
pig-footed bandicoots, DNA from, 41
piopios, DNA from, 41
placental mammals, 42
plague, the
Black Death and, 120–22, 124, 195
plants
ancient DNA, 74–6, 85
Pleistocene epoch, 50
'Pleistocene Park', 63
polymerase chain reaction (PCR), 86–7, 108, 132, 140
Prince Frederick, 156, 157
Prince Philip, 162
Pyramid Valley Swamp, 97, 98

quagga
breeding project, 64–5
extracting DNA from, 39–41
Queen Elizabeth, 162
Queen Victoria, 162

Raoult, Didier, 122–5, 132
Rasputin, Grigory, 147
ratite family, 95, 99
DNA extraction from, 103
sexual dimorphism, 107
Rau, Reinhold, 39–40, 64
religion
early science and, 13
humans, creation of, 11–12

Repin, Vladimir, 62
resins, 77
ribonucleic acid *see* RNA
Ridley, Matt, 89
RNA, 140
Romanovs
  DNA and, 160–2
  fate of, 144, 158
  royal grave, search for, 158–60
  rumoured murder of, 150–1
Ruffman, Alan, 170
Rule, Dr John, 91, 93
Russian Nobility Association, 166
Russian Revolution 1917
  brief history of, 145–51
  effects of, 144
Russian royals
  DNA and, 160–2
  grave, search for, 158–60
Ryabov, Gely, 159

sabre-toothed cats
  DNA from, 41
  Ice Age megafauna, 51
Salo, Wilmar, 131–2
sauropods, 69
Schaafhausen, Hermann, 13–14
Schanzkowska, Franziska, 167–9
Schiefer, Magda, 171
Schweitzer, Dick, 165–7
Schweitzer, Marina, 165, 167
Schweitzer, Mary, 73, 83
science
  development of, 13
  religion and, 13
Scott, Susan, 121–2, 124
sex chromosomes
  birds, 108
  humans, 108
sex determination, human bones, 108
sexual dimorphism, 107
Simon, Antoine, 177–8, 179
Simon, Marie, 177, 179–80
sloth
  DNA from, 41
  extinction, 52
  Ice Age megafauna, 51
Sokolov, Nikolai, 150–1, 152
Solo River, 17
Southland Museum, 103
species, 97–8
Stalin, 158
Steller's sea cows, DNA from, 41
Stone, Anne, 31
Stoneking, Mark, 26–8, 31, 166
Stringer, Chris, 23–4, 28, 32
syphilis, 195

Tasmanian devil, 47, 48–9
Tasmanian Field Naturalists Club, 43
Tasmanian tiger (thylacine), 6
  appearance, 42
  bounty scheme (1888), 43
  bringing back, 41
  cloning, 45–9
  DNA from, 41
  extinction, 37, 41, 44
  genome project, 47
  last, death of, 37, 44
  marsupial, 42
  names, 42
  reported sightings, 45
  scientific name, 42
Taubengberger, Jeffrey, 138–41
Temple of Paris, 176, 181, 182, 191
thylacine *see* Tasmanian tiger
thymine (T), 2, 30, 140
Titanic
  unknown child, 169–72
Triassic period, 68
  amber, 77
Tsar Nicholas II
  Anastasia, daughter of, 143
  DNA identification, 162–4
tuberculosis
  ancient DNA analysis, 195

tuberculosis *continued*
  Columbus and, 112, 114, 126–31
  DNA, 131–3
  epidemics, 112, 114
Tuileries palace, 175, 176
Tyrannosaurus rex, 69, 74

uracil (U), 140

Vaughn, Roscoe, 139
velociraptors, 69
Vereshchagin, Nikolai, 57
Virchow, Rudolf, 15

Williams, William, 94
Wilson, Allan, 26–8, 85
Wolpoff, Milford, 24–5, 28, 32, 33
Woodward, Scott, 80–1, 83, 84
woolly mammoths *see also* mammoths
  ancient DNA, 6
  appearance, 55
  extinction, 37
  Ice Age megafauna, 51
  permafrost, preservation in, 54
woolly rhinoceros
  DNA from, 41
  Ice Age megafauna, 51
  permafrost, preservation in, 54

Yakutsk's Mammoth Museum, 62
Yorkshire Museum, 103
Yucatan Peninsula, 71

Zimov, Sergei, 63
Zoological Society of London, 93